Experiments in dc/ac
Basics to Accompany
dc/ac:
The Basics

Experiments in dc/ac Basics to Accompany
dc/ac: The Basics

Robert L. Boylestad
Gabriel Kousourou

Merrill Publishing Company
A Bell & Howell Information Company
Columbus Toronto London Melbourne

NOTICE TO THE READER

The publisher and the author(s) do not warrant or guarantee any of the products and/or equipment described herein nor has the publisher or the author(s) made any independent analysis in connection with any of the products, equipment, or information used herein. The reader is directed to the manufacturer for any warranty or guarantee for any claim, loss, damages, costs, or expense, arising out of or incurred by the reader in connection with the use or operation of the products and/or equipment.

The reader is expressly advised to adopt all safety precautions that might be indicated by the activities and experiments described herein. The reader assumes all risks in connection with such instructions.

Cover Photo: Jim Osburn, Osburn Photographic Illustration, Inc.

Published by Merrill Publishing Company
A Bell & Howell Information Company
Columbus, Ohio 43216

This book was set in Times Roman and Lubalin Graph.

Administrative Editor: Steve Helba
Production Coordinator: Rex Davidson
Cover Designer: Cathy Watterson
Text Designer: Cynthia Brunk

Library of Congress Catalog Card Number: 88-63593
International Standard Book Number: 0-675-21131-X
Printed in the United States of America
1 2 3 4 5 6 7 8 9—93 92 91 90 89

To our students

Merrill's International Series in Electrical and Electronics Technology

Preface

The importance of a good laboratory experience cannot be overemphasized in the development of the practical knowledge required of a qualified technician. It is usually the only opportunity for a new student of the subject to verify the important concepts introduced in class and to become familiar with the tools of the trade. Most students relish the opportunity to construct, design, or test a physical system, often spending what seems to be endless hours with the theoretical presentation and exercises.

The experiments in this manual progress in the same logical order as in the text, so that the material can be covered in class before it is encountered in the laboratory. Included for each experiment is a résumé that briefly reviews the important concepts to be analyzed. The discussion is brief because it is assumed the material has already been covered in class, and instructors are given an opportunity to introduce their own methods and/or philosophy. Space is provided for most of the calculations and answers to questions, and each page is perforated for easy removal and submission.

The experiments have been carefully tested, reflecting the authors' desire to ensure that the important basic concepts and measurements are clearly and correctly understood. The time required to perform each experiment and respond to the questions is usually about three hours. Of course, the students' preparation before the laboratory exercise will have a direct impact on the required time period. A sufficient number of laboratory experiments have been included to permit a choice of experiments for a two-semester sequence—typically corresponding with a course on dc concepts followed by a course on ac material.

The number of components and instruments required to run each experiment has been kept to an absolute minimum. In fact, the entire set of experiments was chosen to minimize the total list. In addition, the power-level requirements are sufficiently low to permit 1/2-watt (or lower) resistors.

The authors are naturally interested in the response of the users of the manual to the content, length, and effectiveness of the experiments. Please feel free to pass your comments on to the publisher's field representative or main office. We will reply personally to any suggestions received.

We wish to thank Steve Helba and Rex Davidson at Merrill Publishing for their support and guidance in producing this manual. We believe the end product reflects our initial goals, and we sincerely hope it meets the goals of your laboratory experience.

Robert Boylestad
Westport, CT

Gabriel Kousourou
Bayside, NY

Contents

ac EXPERIMENTS

Math Review

dc

The analysis of dc circuits requires that a number of fundamental mathematical operations be performed on an ongoing basis. Most of the operations are typically covered in a secondary-level education; a few others will be introduced in the early sessions of the lecture portion of this course.

The material to follow is a brief review of the important mathematical procedures along with a few examples and exercises. Derivations and detailed explanations are not included but are left as topics for the course syllabus or as research exercises for the student.

It is strongly suggested that the exercises be performed with and without a calculator. Although the modern-day calculator is a marvelous development, the student should understand how to perform the operations in the long-hand fashion. Any student of a technical area should also develop a high level of expertise with the use of the calculator through frequent application with a variety of operations.

FRACTIONS

Addition and Subtraction

Addition and subtraction of fractions require that each term have a common denominator. A direct method of determining the common denominator is simply to multiply the numerator and denominator of each fraction by the denominator of the other fractions. The method of "selecting the least common denominator" will be left for a class lecture or research exercise.

EXAMPLES

$$\frac{1}{2} + \frac{2}{3} = \left(\frac{3}{3}\right)\left(\frac{1}{2}\right) + \left(\frac{2}{2}\right)\left(\frac{2}{3}\right) = \frac{3}{6} + \frac{4}{6} = \frac{7}{6}$$

$$\frac{3}{4} - \frac{2}{5} = \left(\frac{5}{5}\right)\left(\frac{3}{4}\right) - \left(\frac{4}{4}\right)\left(\frac{2}{5}\right) = \frac{15}{20} - \frac{8}{20} = \frac{7}{20}$$

$$\frac{5}{6} + \frac{1}{2} - \frac{3}{4} = \left(\frac{2}{2}\right)\left(\frac{4}{4}\right)\left(\frac{5}{6}\right) + \left(\frac{6}{6}\right)\left(\frac{4}{4}\right)\left(\frac{1}{2}\right) - \left(\frac{6}{6}\right)\left(\frac{2}{2}\right)\left(\frac{3}{4}\right)$$

$$= \frac{40}{48} + \frac{24}{48} - \frac{36}{48} = \frac{28}{48} = \frac{7}{12}$$

(12 being the least common denominator for the three fractions)

Multiplication

The multiplication of fractions is straightforward, with the products of the numerators and denominators formed separately as shown below.

EXAMPLES

$$\left(\frac{1}{3}\right)\left(\frac{4}{5}\right) = \frac{(1)(4)}{(3)(5)} = \frac{4}{15}$$

$$\left(\frac{3}{7}\right)\left(\frac{1}{2}\right)\left(-\frac{4}{5}\right) = \frac{(3)(1)(-4)}{(7)(2)(5)} = -\frac{12}{70} = -\frac{6}{35}$$

Division

Division requires that the denominator be inverted and then multiplied by the numerator.

EXAMPLE

$$\frac{\dfrac{2}{3}}{\dfrac{4}{5}} = \frac{2}{3} \times \frac{5}{4} = \frac{10}{12} = \frac{5}{6}$$

Simplest Form

A fraction can be reduced to simplest form by finding the largest common divisor of both the numerator and denominator. As an illustration,

$$\frac{12}{60} = \frac{12/12}{60/12} = \frac{1}{5}$$

If the largest divisor is not obvious, start with a smaller common divisor, such as shown here:

$$\frac{12/6}{60/6} = \frac{2}{10} = \frac{2/2}{10/2} = \frac{1}{5}$$

Or, if you prefer, use repeated divisions of smaller common divisors, such as 2 or 3:

$$\frac{12/2}{60/2} = \frac{6}{30} = \frac{6/2}{30/2} = \frac{3}{15} = \frac{3/3}{15/3} = \frac{1}{5}$$

EXAMPLES

$$\frac{180}{300} = \frac{180/10}{300/10} = \frac{18}{30} = \frac{18/6}{30/6} = \frac{3}{5}$$

$$\frac{72}{256} = \frac{72/8}{256/8} = \frac{9}{32}$$

or

$$\frac{72}{256} = \frac{72/2}{256/2} = \frac{36}{128} = \frac{36/2}{128/2} = \frac{18}{64} = \frac{18/2}{64/2} = \frac{9}{32}$$

Mixed Numbers

Numbers in the mixed form (whole and fractional part) can be converted to the fractional form by multiplying the whole number by the denominator of the fractional part and adding the numerator.

EXAMPLES

$$3\frac{4}{5} = \frac{(3 \times 5) + 4}{5} = \frac{19}{5}$$

$$60\frac{1}{3} = \frac{(60 \times 3) + 1}{3} = \frac{181}{3}$$

The reverse process is simply a division operation with the remainder left in fractional form.

Conversion to Decimal Form

Fractions can be converted to decimal form by simply performing the indicated division.

EXAMPLES

$$\frac{1}{4} = 4\overline{)1.00}^{\,0.25} = 0.25$$

$$\frac{3}{7} = 7\overline{)3.0000}^{\,0.4285\,\ldots} = 0.4285\ldots$$

$$4\frac{1}{5} = 4 + 5\overline{)1.0}^{\,0.2} = 4.2$$

PROBLEMS

Perform each of the following operations by hand and compare with the calculator solution.

1. $\dfrac{3}{5} + \dfrac{1}{3} =$ _____ (by hand) _____ (calculator)

2. $\dfrac{5}{9} - \dfrac{1}{4} =$ _____ (by hand) _____ (calculator)

3. $\dfrac{9}{10} - \dfrac{1}{2} + \dfrac{1}{4} =$ _____ (by hand) _____ (calculator)

4. $\left(\dfrac{5}{6}\right)\left(\dfrac{9}{13}\right) =$ _____ (by hand) _____ (calculator)

5. $\left(-\dfrac{1}{7}\right)\left(+\dfrac{8}{15}\right)\left(-\dfrac{3}{5}\right) =$ _____ (by hand) _____ (calculator)

6. $\dfrac{5/6}{2/3} =$ _____ (by hand) _____ (calculator)

7. $\dfrac{9/10}{-6/7} =$ _____ (by hand) _____ (calculator)

8. Reduce to simplest form:

     a. $\dfrac{72}{84} =$ _____ (by hand) _____ (calculator)

     b. $\dfrac{144}{384} =$ _____ (by hand) _____ (calculator)

9. Convert to the mixed form:

    a. $\dfrac{16}{7} =$ _____

    b. $\dfrac{320}{9} =$ _____

10. Convert to decimal form:

    a. $\dfrac{3}{8} =$ _____ (by hand) _____ (calculator)

    b. $6\dfrac{5}{6} =$ _____ (by hand) _____ (calculator)

SCIENTIFIC NOTATION

The need to work with very small and very large numbers requires that the use of powers of 10 be appreciated and clearly understood.

The direction of the shift of the decimal point and the number of places the decimal point is moved will determine the resulting power of 10. For instance:

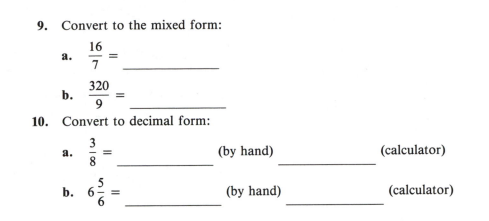

Addition and Subtraction

Addition or subtraction of numbers using scientific notation requires that the power of 10 of each term be the same.

EXAMPLES

$$45,000 + 3000 + 500 = 45 \times 10^3 + 3 \times 10^3 + 0.5 \times 10^3 = 48.5 \times 10^3$$

$$0.02 - 0.003 + 0.0004 = 200 \times 10^{-4} - 30 \times 10^{-4} + 4 \times 10^{-4}$$

$$= (200 - 30 + 4) \times 10^{-4} = 174 \times 10^{-4}$$

Multiplication

Multiplication using powers of 10 employs the following equation, where a and b can be any positive or negative number:

$$\boxed{10^a \times 10^b = 10^{a+b}}$$ **(M.1)**

EXAMPLES

$$(100)(5000) = (10^2)(5 \times 10^3) = 5 \times 10^2 \times 10^3 = 5 \times 10^{2+3} = 5 \times 10^5$$

$$(200)(0.0004) = (2 \times 10^2)(4 \times 10^{-4}) = (2)(4)(10^2)(10^{-4}) = 8 \times 10^{2-4} = 8 \times 10^{-2}$$

Note in the above examples that the operations with powers of 10 can be separated from the integer values.

Division

Division employs the following equation, where a and b can again be any positive or negative number:

$$\boxed{\frac{10^a}{10^b} = 10^{a-b}} \qquad \text{(M.2)}$$

EXAMPLES

$$\frac{320,000}{4000} = \frac{32 \times 10^4}{4 \times 10^3} = \frac{32}{4} \times 10^{4-3} = 8 \times 10^1 = 80$$

$$\frac{1600}{0.0008} \quad \frac{16 \times 10^2}{8 \times 10^{-4}} = 2 \times 10^{2-(-4)} = 2 \times 10^{2+4} = 2 \times 10^6$$

Note in the last example the importance of carrying the proper sign through Eq. (M.2).

Powers

Powers of powers of 10 can be determined using the following equation, where a and b can be any positive or negative number:

$$\boxed{(10^a)^b = 10^{ab}} \qquad \text{(M.3)}$$

EXAMPLES

$$(1000)^4 = (10^3)^4 = 10^{12}$$

$$(0.002)^3 = (2 \times 10^{-3})^3 = (2)^3 \times (10^{-3})^3 = 8 \times 10^{-9}$$

PROBLEMS

Using powers of 10, perform the following operations by hand, and then compare with the calculator solution.

11. $5,800,000 + 450,000 + 2000 =$ _____ (by hand) _____ (calculator)

12. $0.04 + 0.008 - 0.3 =$ _____ (by hand) _____ (calculator)

13. $2400 + 0.05 \times 10^3 - 40,000 \times 10^{-3} =$ _____ (by hand)
_____ (calculator)

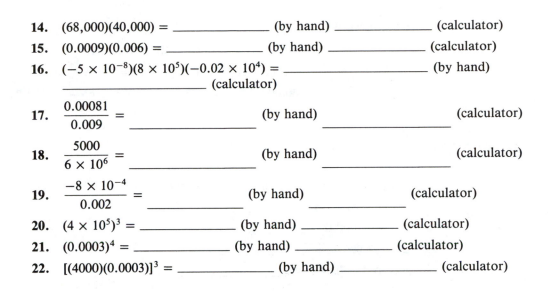

14. (68,000)(40,000) = _____ (by hand) _____ (calculator)

15. (0.0009)(0.006) = _____ (by hand) _____ (calculator)

16. $(-5 \times 10^{-8})(8 \times 10^{5})(-0.02 \times 10^{4})$ = _____ (by hand) _____ (calculator)

17. $\dfrac{0.00081}{0.009}$ = _____ (by hand) _____ (calculator)

18. $\dfrac{5000}{6 \times 10^{6}}$ = _____ (by hand) _____ (calculator)

19. $\dfrac{-8 \times 10^{-4}}{0.002}$ = _____ (by hand) _____ (calculator)

20. $(4 \times 10^{5})^{3}$ = _____ (by hand) _____ (calculator)

21. $(0.0003)^{4}$ = _____ (by hand) _____ (calculator)

22. $[(4000)(0.0003)]^{3}$ = _____ (by hand) _____ (calculator)

PREFIXES

The frequent use of some powers of 10 has resulted in their being assigned abbreviations that can be applied as prefixes to a numerical value in order to quickly identify its relative magnitude.

The following is a list of the most frequently used prefixes in electrical and electronic technology:

$$10^{6} = \text{mega (M)}$$
$$10^{3} = \text{kilo (k)}$$
$$10^{-3} = \text{milli (m)}$$
$$10^{-6} = \text{micro } (\mu)$$
$$10^{-9} = \text{nano (n)}$$
$$10^{-12} = \text{pico (p)}$$

EXAMPLES

$$6,000,000 \ \Omega = 6 \text{ megohms} = 6 \text{ M}\Omega$$
$$0.04 \text{ A} = 40 \times 10^{-3} \text{ A} = 40 \text{ mA}$$
$$0.00005 \text{ V} = 50 \times 10^{-6} \text{ V} = 50 \ \mu\text{V}$$

When converting from one form to another, be aware of the relative magnitude of the quantity before you start to provide a check once the conversion is complete. Starting out with a relatively small quantity and ending up with a result of large relative magnitude clearly indicates an error in the conversion process. In many cases the best method may be to return to the decimal form before establishing the new form for the quantity.

EXAMPLES

$$0.008 \text{ M}\Omega = 0.008 \times 10^6 \text{ }\Omega = 8 \times 10^3 \text{ }\Omega = 8 \text{ k}\Omega$$
$$5600 \text{ mA} = 5600 \times 10^{-3} \text{ A} = 5.6 \text{ A}$$
$$20,000 \text{ mV} = 20,000 \times 10^{-3} \text{ V} = 20 \text{ V} = 0.02 \text{ kV}$$

PROBLEMS

Apply the most appropriate prefix to each of the following quantities.

23. $0.00006 \text{ A} = $ _____

24. $504,000 \text{ }\Omega = $ _____

25. $32,000 \text{ V} = $ _____

26. $0.000000009 \text{ A} = $ _____

Perform the following conversions.

27. $35 \text{ mA} = $ _____ A

28. $0.005 \text{ kV} = $ _____ V = _____ mV

29. $8,000,000 \text{ }\Omega = $ _____ MΩ = _____ kΩ

30. $4000 \text{ pF} = $ _____ nF = _____ μF

OTHER FUNCTIONS

The student will encounter a variety of other mathematical functions in the study of electrical and electronic systems. A few will now be examined to ensure their correct evaluation when the need arises.

Square and Cubic Roots of a Number

The calculator will always be used to determine the square and cubic roots of a number. In each case, the power of 10 must be divisible by 2 or 3, respectively, or the resulting power of 10 will not be a whole number.

EXAMPLES

$$\sqrt{200} = (200)^{1/2} = (2 \times 10^2)^{1/2} = (2)^{1/2} \times (10^2)^{1/2} \cong 1.414 \times 10^1 = 14.14$$
$$\sqrt{0.004} = (0.004)^{1/2} = (40 \times 10^{-4})^{1/2} = (40)^{1/2} \times (10^{-4})^{1/2}$$
$$= 6.325 \times 10^{-2} = 0.06325$$
$$\sqrt[3]{5000} = (5000)^{1/3} = (5 \times 10^3)^{1/3} = (5)^{1/3} \times 10^1 \cong 1.71 \times 10^1 = 17.1$$

Some calculators provide the $\sqrt[3]{}$ function while others require that you use the $\sqrt[x]{y}$ function. For some calculators, the y^x function is all that is available, requiring that the y value be entered first, followed by the y^x function and then the power ($x = 1/3 = 0.3333$).

PROBLEMS

31. $\sqrt{6.4} =$ _____

32. $\sqrt[3]{3000} =$ _____

33. $\sqrt{0.00005} =$ _____

EXPONENTIAL FUNCTIONS

The exponential functions e^x and e^{-x} will appear frequently in the analysis of R-C and R-L networks with switched dc inputs (or square-wave inputs). On most calculators, only e^x is provided, requiring that the user insert the negative sign when necessary.

Keep in mind that e^x is equivalent to $(2.71828 \ldots)^x$ or a number to a power x. For any positive values of x greater than 1, the result is a magnitude greater than $2.71828 \ldots$. In other words, for increasing values of x, the magnitude of e^x will increase rapidly. For $x = 0$, $e^0 = 1$, and for increasing negative values of x, e^{-x} will become increasingly smaller.

When inserting the negative sign for e^{-x} functions, be sure to use the proper key to enter the negative sign. It is not always the key used for addition.

EXAMPLES

$$e^{+2} \cong 7.3890$$

$$e^{+10} \cong 22026.47$$

$$e^{-2} \cong 0.13534$$

$$e^{-10} \cong 0.0000454$$

PROBLEMS

34. $e^{+4.2} =$ _____

35. $e^{+0.5} =$ _____

36. $e^{-0.02} =$ _____

ALGEBRAIC MANIPULATIONS

The following is a brief review of some basic algebraic manipulations that must be performed in the analysis of dc circuits. It is assumed that a supporting math course will expand on the coverage provided here.

EXAMPLES

(a) Given $v = \dfrac{d}{t}$, solve for d and t.

Both sides of the equation must first be multiplied by t:

$$(t)(v) = \left(\frac{d}{\cancel{t}}\right)(\cancel{t})$$

resulting in $d = vt$.

Dividing both sides of $d = vt$ by v yields

$$\frac{d}{v} = \frac{\cancel{v}t}{\cancel{v}}$$

and

$$t = \frac{d}{v}$$

In other words, the proper choice of multiplying factors for both sides of the equation will result in an equation for the desired quantity.

(b) Given $R_1 + 4 = 3R_1$, solve for R_1.

In this case, $-R_1$ is added to both sides, resulting in

$$(R_1 + 4) - R_1 = 3R_1 - R_1$$

and

$$4 = 2R_1$$

with

$$R_1 = \frac{4}{2} = 2 \; \Omega$$

(c) Given $\frac{1}{R_1} = \frac{1}{3R_1} + \frac{1}{6}$, solve for R_1.

Multiplying both sides of the equation by R_1 results in

$$\frac{\cancel{R_1}}{\cancel{R_1}} = \frac{\cancel{R_1}}{3\cancel{R_1}} + \frac{R_1}{6}$$

or

$$1 = \frac{1}{3} + \frac{R_1}{6}$$

Subtracting 1/3 from both sides gives

$$1 - \frac{1}{3} = \frac{1}{3} - \frac{1}{3} + \frac{R_1}{6}$$

and

$$\frac{2}{3} = \frac{R_1}{6}$$

with

$$R_1 = \frac{(6)(2)}{3} = \frac{12}{3} = 4 \; \Omega$$

PROBLEMS

37. Given $I = \dfrac{E}{R}$, solve for R and E.

$R =$ _____ , $E =$ _____

38. Given $30I = 5I + 5$, solve for I.

$I =$ _____

39. Given $\dfrac{1}{R_T} = \dfrac{1}{R_1} + \dfrac{1}{4R_1}$, solve for R_T in terms of R_1.

$R_T =$ _____

40. Given $F = \dfrac{kQ_1Q_2}{r^2}$, solve for Q_1 and r.

$Q_1 =$ _____ , $r =$ _____

Resistors and the Color Code

OBJECT

To become familiar with the use of the ohmmeter and the resistor color code.

EQUIPMENT REQUIRED

Resistors

1 — 1-MΩ, 1-W carbon resistor

1 — 1-MΩ, 2-W carbon resistor

1 — 91-Ω, 220-Ω, 2.2-kΩ, 10-kΩ, 1-MΩ, 1/2-W carbon resistors

Instruments

1 — VOM (Volt-Ohm-Milliammeter)

1 — DMM (Digital Multimeter)

EQUIPMENT ISSUED

TABLE 1.1

Item	Manufacturer and Model No.	Laboratory Serial No.
VOM		
DMM		

DESCRIPTION OF EQUIPMENT

Both the VOM and the DMM will be used in this experiment to measure the resistance of the provided resistors. The VOM employs an analog scale to read resistance, voltage, and current, while the DMM has a digital display. An analog scale is a continuous scale, requiring that the user be able to interpret the location of the pointer using the scale divisions provided. For resistance measurements, the analog scale is also nonlinear, resulting in smaller distances between increasing values of resistance. Note on the VOM scale the relatively large distance between 1 Ω and 10 Ω and the smaller distance between 100 Ω and 1000 Ω. The digital display provides a numerical value with the accuracy determined by the chosen scale. For many years the analog meter was the instrument employed throughout the industry. In recent years the digital meter has grown in popularity, making it important that the graduate of any technical program be adept at using both types of meters.

RÉSUMÉ OF THEORY

In this experiment the resistance of a series of 1/2-W carbon resistors will first be determined from the color code and then compared with the measured value using both the VOM and the DMM. The two meters are applied to ensure familiarity with reading both an analog and a digital scale.

The procedure for determining the resistance of a color-coded resistor is described in Appendix I with a listing of the numerical value associated with each color. The first two bands (those closest to the end of the resistor) determine the first two digits of the resistor value, while the third band determines the power of the power-of-10 multiplier (actually the number of zeros to follow the first two digits). The fourth band is the percent tolerance for the chosen resistor (see Fig. 1.1).

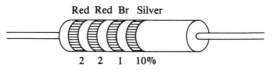

Red Red Br Silver

2   2   1   10%

FIG. 1.1   220 ± 10% = 220 ± 22 = 198 Ω to 242 Ω

For increasing wattage ratings, the size of the carbon resistor will increase to provide the "body" required to dissipate the resulting heating effects.

Resistance is *never* measured by an ohmmeter in a *live* network, due to the possibility of damaging the meter with excessively high currents and obtaining readings that have no meaning. It is usually best to remove the resistor from the circuit before measuring its resistance to ensure that the measured value does not include the effect of other resistors in the system. However, if this maneuver is undesirable or impossible, make sure that one end of the resistor is not connected to any other element.

The use of the VOM and DMM will be described early in the laboratory session. For the VOM, always reset the zero-adjust whenever you change scales. In addition, always choose the dial setting (R × 1, R × 10, and so on) that will place the pointer in the region of the scale that will give the best reading. Finally, do not forget to multiply the reading by the proper multiplying factor. For the DMM, remember that any scale marked "kΩ" will be reading in kilohms, and any "MΩ" scale in megohms. For instance, a 91-Ω resistor may read 90.7 Ω on a 200-Ω scale, 0.091 on a 2-kΩ scale, and 0.000 on a 2-MΩ scale. There is no zero-adjust on a DMM meter, but make sure that $R = 0 \ \Omega$ when the leads are touching or an adjustment internal to the meter may have to be made. Any resistance above the maximum for a chosen scale will result in an O.L. indication.

One last comment is in order. Remember that there is no polarity to resistance measurements. Either lead of the meter can be placed on either end of the resistor—the resistance will be the same. Polarities will become important, however, when we measure voltages and currents in the experiments to follow.

PROCEDURE

The purpose of this first experiment is to acquaint you with the equipment, so *do not rush*. Learn how to read the meter scales accurately, and take your data carefully. If you are uncertain about anything, do not hesitate to ask your instructor.

Part 1 Body Size

(a) In the space below, draw the physical sizes of 1/2-W, 1-W, and 2-W, 1-MΩ carbon resistors (color bands = brown, black, green). Note in particular that the resistance of each is the same but that the size increases with wattage rating.

(b) How much larger (approximately) is the 1-W resistor than the 1/2-W resistor? Is the ratio the same for the 2-W resistor as compared with the 1-W resistor?

Part 2 Color Code

(a) Complete Table 1.2 for each of the provided 1/2-W resistors. Examine the resistors in any order, but remember that resistor 1 in Table 1.2 is resistor 1 in the other tables to follow. An example is provided to define the procedure.

TABLE 1.2

Resistor	Color Bands — Color				Color Bands — Numerical Value				Nominal Resistance
	1	2	3	4	1	2	3	4	
Sample	Red	Red	Black	Gold	2	2	0	5%	22 Ω
1									
2									
3									
4									
5									

The percent tolerance is used to determine the range of resistance levels within which the manufacturer guarantees the resistor will fall. It is determined by first taking the percent tolerance and multiplying by the nominal resistance level. For the example in Table 1.2, the resulting resistance level

$$(5\%)(22\ \Omega) = (0.05)(22\ \Omega) = 1.1\ \Omega$$

is added to and subtracted from the nominal value to determine the range as follows:

$$\text{Maximum value} = 22\ \Omega + 1.1\ \Omega = 23.1\ \Omega$$
$$\text{Minimum value} = 22\ \Omega - 1.1\ \Omega = 20.9\ \Omega$$

as shown in Table 1.3.

(b) Complete Table 1.3 for each resistor in Table 1.2.

TABLE 1.3

Resistor	Maximum Resistance	Minimum Resistance
Sample	23.1 Ω	20.9 Ω
1		
2		
3		
4		
5		

The resistance levels will now be read using the VOM and DMM and inserted in Table 1.4. The percent difference will be determined using the following equation:

$$\% \text{ Difference} = \frac{|\text{Measured} - \text{Nominal}|}{\text{Nominal}} \times 100\%$$

For the sample and the DMM reading:

$$\% \text{ Difference} = \frac{|22.9 \ \Omega - 22 \ \Omega|}{22 \ \Omega} \times 100\%$$

$$= \frac{0.9}{22} \times 100\% = \underline{\underline{4.09\%}}$$

TABLE 1.4

Resistor	Measured Value— VOM	Falls Within Specified Tolerance (Yes/No)	Measured Value— DMM	Falls Within Specified Tolerance (Yes/No)	% Difference (Using DMM Reading)
Sample	23 Ω	Yes	22.9 Ω	Yes	4.09
1					
2					
3					
4					
5					

Part 3   Body Resistance

Guess the resistance of your body between your hands and record the value in Table 1.5. Measure the resistance with the DMM (or VOM) by firmly holding one lead in each hand (wetting your fingers will improve the reading), and record in the same table. If 10 mA are "lethal," what voltage ($V = IR$) would be required to produce the current through your body? Again, record in Table 1.5.

TABLE 1.5

Guessed body resistance	
Measured body resistance	
Lethal voltage	

Calculations:

Part 4   Meter Resistance

(a) Voltmeters Ideally, the internal resistance of the DMM and VOM should be infinite (like an open circuit) when voltages in a network are being measured to ensure that the meter does not alter the normal behavior of the network.

For the VOM, there is an ohm/volt (Ω/V) rating written on the bottom of the scale that permits determination of the internal resistance of each scale of the meter when used as a voltmeter. For the 260 Simpson VOM, the ohm/volt rating is 20,000 Ω/V. The internal resistance of each setting can then be calculated by multiplying the maximum voltage reading of a scale by the ohm/volt rating. For instance, the 10-V scale will have (10 V)(20,000 Ω/V) = 200 kΩ, while the 250-V scale will have (250 V)(20,000 Ω/V) = 5 MΩ. No matter what voltage level is measured on each scale, the internal resistance will remain the same.

Using the DMM as an ohmmeter, measure the resistance of each voltage scale of the VOM and record in Table 1.6. Insert the maximum voltage levels of your meter if different from those appearing in the table.

TABLE 1.6

VOM	Scales* of Your VOM	Ω/V Rating	Calculated Resistance	Measured Resistance	% Difference — $\dfrac{\lvert \text{Meas.} - \text{Calc.} \rvert}{\text{Calc.}} \times 100\%$
2.5 V					
10 V					
50 V					
250 V					
500 V					

*If different from those in the first column.

Most DMMs have the same internal resistance for each scale. In other words, the internal resistance of a typical meter will be 10 MΩ whether the 2-V or the 200-V scale is used.

Use the VOM to measure the internal resistance of each scale of the DMM and complete Table 1.7. Make sure the DMM is on.

TABLE 1.7

DMM	Your DMM* (Insert Scale Settings)	Specified Internal Resistance	Measured Resistance
200 mV			
2 V			
20 V			
200 V			
1500 V			

*If different from those in the first column.

In general, which meter would appear to disturb the network less?

(b) Ammeters The ammeter is also an instrument that when inserted in a network should not adversely affect the normal current levels. However, since it is placed in series with the branch in which the current is being measured, it should ideally have a resistance of zero ohms (or as small as possible).

Using the DMM as an ohmmeter, measure the resistance of each current scale of the VOM and record in Table 1.8. Insert the maximum current levels of your meter if different from those appearing in the table.

TABLE 1.8

VOM	Your VOM* (Insert Scale Settings)	Measured Resistance
1 mA		
10 mA		
100 mA		
500 mA		

*If different from those in the first column.

Using the VOM as an ohmmeter, measure the resistance of each current scale of the DMM and record in Table 1.9. As above, insert the maximum current levels of your meter if different from those appearing in the table.

TABLE 1.9

DMM	Your DMM* (Insert Scale Settings)	Measured Resistance
2 mA		
20 mA		
200 mA		
2 A		

*If different from those in the first column.

Which meter would appear to disturb the network the least?

PROBLEMS

1.   What are the ohmic values and tolerances of the following carbon resistors?

Color Bands—Color				Numerical Value	Tolerance
1	2	3	4		
Brown	Black	Blue	Gold		
Yellow	Violet	Orange	Gold		
Brown	Gray	Gold	None		
Red	Yellow	Silver	Gold		
Green	Brown	Green	Silver		
Green	Blue	Black	None		

2.   For the VOM, which region of the scale normally provides the best readings? Why?

Ohm's Law

OBJECT

*To become familiar with the use of the
laboratory dc supply and with measuring
voltages and currents*

EQUIPMENT REQUIRED

Resistors

$1-91\text{-}\Omega$, 1000-Ω, 3300-Ω, 10,000-Ω, 1-MΩ, 1/2-W

Instruments

1—DMM (or VOM)

1—dc Power supply

EQUIPMENT ISSUED

TABLE 2.1

Item	Manufacturer and Model No.	Laboratory Serial No.
DMM		
VOM		
Power supply		

TABLE 2.2

Resistors	
Nominal (Color-Coded) Value	Measured Value
1,000 Ω	
10,000 Ω	

RÉSUMÉ OF THEORY

In any active circuit there must be a source of power. In the laboratory, it is convenient to use a source that requires a minimum of maintenance and, more important, whose output voltage can be varied easily. Power supplies are rated as to maximum voltage and current output. For example, a supply rated 0–40 V at 500 mA will provide a maximum voltage of 40 V and a maximum current of 500 mA at any voltage.

Most dc power supplies have three terminals, labeled as shown in Fig. 2.1. The

$A \oplus$
$\bigcirc$
$B \ominus$

FIG. 2.1

three terminals permit the establishment of a positive or negative voltage, which can be grounded or ungrounded. The variable voltage is available only between terminals A and B. Both A and B must be part of any connection scheme. If only terminals A and B are employed as shown in Fig. 2.2, the supply is considered "floating" and not connected to the common ground of the network. For common ground and safety reasons, the supply is normally grounded as shown in Fig. 2.3 for a positive voltage, and as in Fig. 2.4 for a negative voltage.

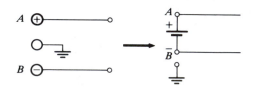

FIG. 2.2

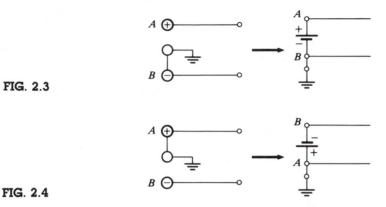

FIG. 2.3

FIG. 2.4

When measuring voltage levels, make sure the voltmeter is connected in parallel (across) the element being measured as shown in Fig. 2.5. In addition, recognize that if the leads are connected as shown in the figure, the reading will be up-scale and positive. If the meter were hooked up in the reverse manner, a negative (down-scale, below-zero) reading would result. The voltmeter is therefore an excellent instrument for not only measuring the voltage level but also determining the polarity. Since the meter is always placed in parallel with the element, there is no need to disturb the network when the measurement is made.

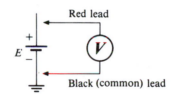

FIG. 2.5

Ammeters are always connected in series with the branch in which the current is being measured as shown in Fig. 2.6, normally requiring that the branch be opened and the meter inserted. Ammeters also have polarity markings to indicate the manner in which they should be connected to obtain an up-scale reading. Since the current I of Fig. 2.6 would establish a voltage drop across the ammeter as illustrated, the reading of the ammeter will be up-scale and positive. If the meter were hooked up in the reverse manner, the reading would be negative or down-scale. In other words, simply reversing the leads will change a below-zero indication to an up-scale reading.

FIG. 2.6

Until you become familiar with the use of the ammeter, draw in the ammeter in the network with the polarities determined by the current direction. It is then easier to ensure that the meter is connected properly to the surrounding elements. This process will be demonstrated in more detail in a later experiment.

For both the voltmeter and the ammeter, always start with the higher ranges and work down to the operating level to avoid damaging the instrument. When the VOM and DMM are returned to the stockroom, be sure the VOM is on the highest voltage scale and the DMM is in the off position.

The voltage across and the current through a resistor can be used to determine its resistance using Ohm's law in the following form:

$$R = \frac{V}{I}$$

(2.1)

The magnitude of R will be determined by the units of measure for V and I.

PROCEDURE

The purpose of this laboratory exercise is to acquaint you with the equipment, so *do not rush*. If you are a member of a squad, don't let one individual make all the measurements. You must become comfortable with the instruments if you expect to perform your future job function in a professional manner. Read the instruments carefully. The more accurate a reading, the more accurate the results obtained. One final word of caution. For obvious reasons, *do not make network changes with the power on!* If you have any questions about the procedure, be sure to contact your instructor.

Part 1

Hook up the DMM (or VOM) to the power supply as shown in Fig. 2.7.

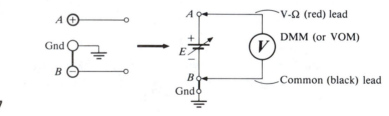

FIG. 2.7

Using the DMM, adjust the power supply to the voltages listed in Table 2.3. Read the voltage V_{AB} at each level using the DMM and VOM.

TABLE 2.3

V_{AB}	DMM	VOM
2 V		
4 V		
6 V		
8 V		
10 V		

Comment on the readings obtained with each meter. Which was easier to use? Which appeared to be more accurate? Answer the questions in some detail.

Part 2 Ohm's Law

(a) Construct the circuit of Fig. 2.8

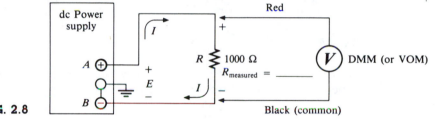

FIG. 2.8

Insert the measured value of R from Table 2.2 and use it for all calculations. Note that the voltmeter is placed in parallel with the resistor with the positive (red) lead connected to the point of higher potential as established by the indicated current direction.

Adjust the power supply until $E = V = 2$ V using the voltmeter. Then turn off the power and construct the circuit of Fig. 2.9 to measure the current I. Note the manner in which the meter is connected to ensure that the meter polarities match those established by the current direction. Record the value of I in Table 2.4.

TABLE 2.4

	$R = 1\ \mathrm{k\Omega}$	
$E = V$	I	$R = \dfrac{V}{I}$
2 V		
4 V		
6 V		
8 V		
10 V		

(b) Repeat the above process for a range of values for E from 2 V to 10 V in 2-V steps and insert the measured value of I for each step in Table 2.4.

(c) Calculate the resistance using the supply voltage and measured current and Ohm's Law and complete the second column. Show all calculations below:

DMM (or VOM) — mA range

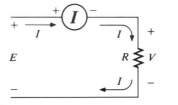

FIG. 2.9

(d) How do the calculated values of R in the second column compare with the nameplate value of the resistor (1 kΩ)?

(e) How do the calculated values of R in the second column compare with the measured value (by an ohmmeter) appearing in Fig. 2.8?

(f) Plot the data from Table 2.4 on Graph 2.1 using the scales provided.

(g) The slope of the curve is related to the resistance by

$$\boxed{\text{Slope} = m = \frac{1}{R}} \qquad (2.2)$$

In other words, the less the resistance the steeper the slope and vice versa. The resistance R can be determined from the graph using the equation $R = V/I$ at any point in the graph.

Determine the resistance of the resistor plotted on Graph 2.1 at $I = 2.6$ mA and at $V = 8.2$ V.

$I = 2.6$ mA:

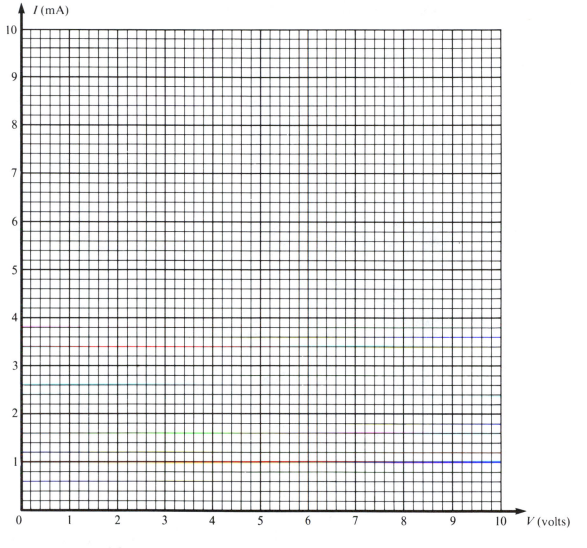

GRAPH 2.1

$V = 8.2$ V:

(h) The resistance can also be determined from the equation

$$R = \frac{\Delta V}{\Delta I}$$ (2.3)

where ΔI is the corresponding change in current due to a change in voltage ΔV or vice versa.

Determine the resistance of the resistor plotted on Graph 2.1 using a $\Delta V = 6\ V - 2\ V = 4\ V$

Using the resistance level just calculated, find the slope m of the curve using Eq. (2.2).

(i) Complete Table 2.5 for $R = 10\ k\Omega$ using the same procedure described for parts 2(a) and 2(b)

TABLE 2.5

		$R = 10\ k\Omega$	
E	I		$R = \dfrac{V}{I}$
2 V			
4 V			
6 V			
8 V			
10 V			

(j) Calculate the resistance using the supply voltage and measured current and Ohm's Law and complete the second column of Table 2.5. Show all calculations below:

(k) How do the calculated values of R in the second column of Table 2.5 compare with the nameplate value of the resistor (10 kΩ)?

(**l**) How do the calculated values of R in the second column of Table 2.5 compare with the measured value (by an ohmmeter) of the 10-kΩ resistor?

(**m**) Plot the data from Table 2.5 on Graph 2.1 using the scales provided.

(**n**) How would you compare the slopes of the curves?

(**o**) Do the results of part 2(n) substantiate Eq. (2.2)?

(**p**) Determine the resistance at $V = 5.6$ V and compare with the measured value.

$R = $ _____

(**q**) Determine the resistance using a $\Delta V = 6$ V $- 2$ V $= 4$ V and Eq. (2.3) and compare the results with the measured value.

$R = $ _____

PROBLEMS

1. Plot the linear curve for a 100-Ω resistor on Graph 2.1 and comment on the slope compared with those for the other resistors.

2. Sketch the proper hookup for two parallel grounded supplies providing 10 V at twice the current rating of either supply.

3. Why does the equation $R = V/I$ provide the same result as the equation $R = \Delta V/\Delta I$ for any level of V or ΔV for the fixed carbon composition resistors?

Series Resistance

OBJECT

To determine the total resistance of series dc circuits.

EQUIPMENT REQUIRED

Resistors

1 — 220-Ω, 330-Ω, 1-kΩ, 1-MΩ
3 — 100-Ω

Instruments

1 — DMM (or VOM)
1 — dc Power supply

EQUIPMENT ISSUED

TABLE 3.1

Item	Manufacturer and Model No.	Laboratory Serial No.
DMM (or VOM)		
Power supply		

TABLE 3.2

Resistors	
Nominal Value	Measured Value
100 Ω	
100 Ω	
100 Ω	
220 Ω	
330 Ω	
1 kΩ	
1 MΩ	

RÉSUMÉ OF THEORY

The total resistance R_T of a series circuit is the sum of the individual resistances. For Fig. 3.1,

$$R_T = R_1 + R_2 + R_3 \qquad (3.1)$$

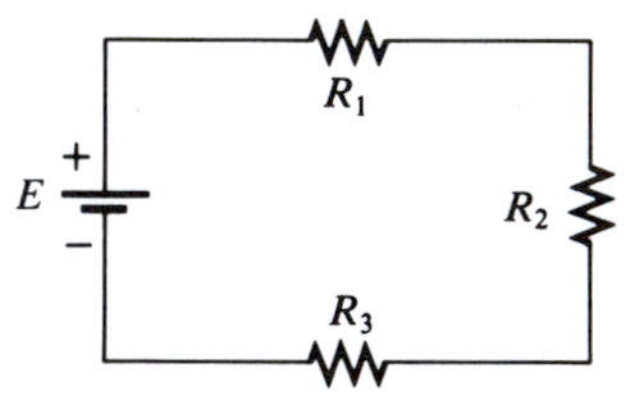

FIG. 3.1

For equal series resistors, the total resistance is equal to the resistance of one resistor times the number in series. That is,

$$R_T = NR \qquad (3.2)$$

For an unknown system, such as that appearing in Fig. 3.2, the total resistance can be determined by the following form of Ohm's law:

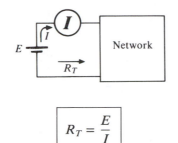

FIG. 3.2

$$R_T = \frac{E}{I}$$ (3.3)

That is, for an applied voltage E the current is measured and Eq. (3.3) applied to determine the resistance of the network.

For resistors of difference magnitudes in series, the total resistance is determined primarily by the largest resistors. In some instances it is very close to the value of one resistor if that resistor is much larger than the others.

PROCEDURE

Part 1   Two Series Resistors

(a) Construct the circuit of Fig. 3.3.

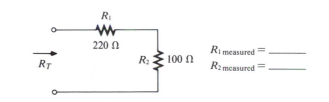

FIG. 3.3

In the spaces provided, insert the measured value of each resistor as determined using an ohmmeter.

(b) Calculate the total resistance using the measured resistor values.

$R_T =$ _____

(c) Measure the total resistance using the ohmmeter section of your DMM or VOM.

$R_T =$ _____

How do the calculated and measured values compare? Calculate the percent difference from

$$\% \text{ Difference} = \frac{|\text{Measured} - \text{Calculated}|}{\text{Calculated}} \times 100\% \qquad (3.4)$$

% Difference = _____

(d) Apply 10 V to the circuit and insert an ammeter as shown in Fig. 3.4. Set the supply voltage with the voltmeter section of your meter and then turn off the supply. Insert the meter in the ammeter mode as shown and turn on the supply. Record the ammeter reading below.

$I = $ _____

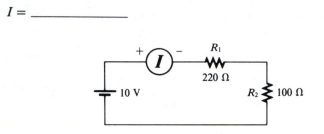

FIG. 3.4

(e) Using the supply voltage and ammeter reading of part 1(d), calculate the total resistance using Eq. (3.3).

$R_T = $ _____

How do the results of this part and part 1(c) compare?

Part 2   Three Series Resistors

(a) Construct the circuit of Fig. 3.5. Insert the measured values of the resistors in the spaces provided.

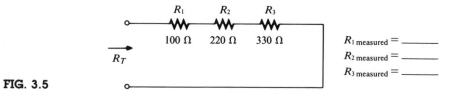

FIG. 3.5

(b) Calculate the total resistance using the measured resistor values.

$R_T =$ _____

(c) Measure the total resistance using the ohmmeter section of your DMM or VOM.

$R_T =$ _____

How do the calculated and measured values compare? Calculate the percent difference using Eq. (3.4).

% Difference = _____

(d) Apply 10 V and an ammeter in the same manner described in part 1(d) and record the source current below.

$I =$ _____

(e) Using the supply voltage and ammeter reading of part 2(d), calculate the total resistance using Eq. (3.3).

$R_T =$ _____

How do the results of this part and part 2(c) compare?

Part 3   Equal Series Resistors

(a) Construct the circuit of Fig. 3.6. Insert the measured values of the resistors in the spaces provided.

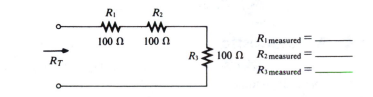

FIG. 3.6

(b) Using the nominal resistor values (all 100 Ω), calculate the total resistance using Eq. (3.2).

$R_T =$ _____

(c) Measure the total resistance using the ohmmeter section of your DMM or VOM.

$R_T =$ _____

How do the calculated and measured values compare? Calculate the percent difference using Eq. (3.4).

% Difference = _____

(d) Calculate the total resistance using the measured values of each resistor.

$R_T =$ _____

How does this calculated value compare with the measured value of part 3(c)? What is the percent difference?

% Difference = _____

Part 4   A Composite

(a) Construct the circuit of Fig. 3.7. Insert the measured resistor values in the spaces provided.

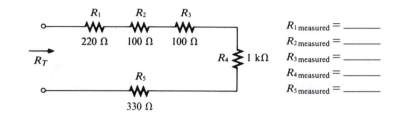

FIG. 3.7

(b) Calculate the total resistance using the measured values.

$R_T = $ _____

(c) Measure the total resistance and find the percent difference between the calculated and measured values.

$R_T = $ _____

% Difference = _____

(d) Apply 10 V and measure the source current I. Then calculate the total resistance and compare with the measured value of part 4(c).

$I = $ _____

$R_T = $ _____

% Difference = _____

Part 5　Different Levels of Resistance

(a) Construct the circuit of Fig. 3.8. Insert the measured resistor values in the spaces provided.

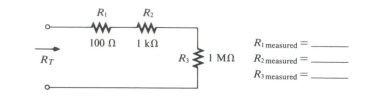

FIG. 3.8

(b) Using the measured values, calculate the total resistance of the circuit.

$R_T =$ _____

(c) Using an ohmmeter, measure the total resistance.

$R_T =$ _____

Calculate the percent difference between the measured and calculated values.

% Difference = _____

(d) Calculate the total resistance using measured values if the resistance R_1 is ignored.

$R_T =$ _____

What is the percent difference between this R_T and the measured value of part 5(c)?

% Difference = _____

(e) What is the total resistance using the measured value if both the resistances R_1 and R_2 are ignored.

$R_T =$ _____

What is the percent difference between this R_T and the measured value of part 5(c)?

% Difference = _____

(f) Noting the percent differences of parts 5(d) and 5(e), what approximations do you think can be made when calculating the total resistance of Fig. 3.8?

Series dc Circuits

OBJECT

To investigate the characteristics of a series dc circuit

EQUIPMENT REQUIRED

Resistors

1 — 330-Ω, 470-Ω

2 — 220-Ω

Instruments

1 — DMM (or VOM)

1 — dc Power supply

EQUIPMENT ISSUED

TABLE 4.1

Item	Manufacturer and Model No.	Laboratory Serial No.
DMM (or VOM)		
Power supply		

TABLE 4.2

Resistors	
Nominal Value	**Measured Value**
220 Ω	
220 Ω	
330 Ω	
470 Ω	

RÉSUMÉ OF THEORY

In a series circuit, the current is the same through all of the circuit elements, as shown by the current I in Fig. 4.1.

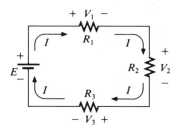

FIG. 4.1

By Ohm's law, the current I is $I = E/R_T$, where E is the impressed voltage and R_T the total resistance. Applying Kirchhoff's voltage law around the closed loop of Fig. 4.1, we find

$$\boxed{E = V_1 + V_2 + V_3}$$

(4.1)

with $V_1 = IR_1$, $V_2 = IR_2$, and $V_3 = IR_3$.

The voltage divider rule states that the voltage across an element or across a series combination of elements in a series circuit is equal to the resistance of the element or series combination of elements divided by the total resistance of the series circuit and multiplied by the total impressed voltage. For the elements of Fig. 4.1,

$$V_1 = \frac{R_1 E}{R_T} = \frac{R_1 E}{R_1 + R_2 + R_3} \quad , \quad V_2 = \frac{R_2 E}{R_T} \quad , \quad V_3 = \frac{R_3 E}{R_T}$$

PROCEDURE

Part 1   Series Resistance Measurements

(a) Construct the circuit shown in Fig. 4.2.

Record the measured value of each resistor in the place indicated. Use the DMM (or VOM) to set $E = 20$ V.

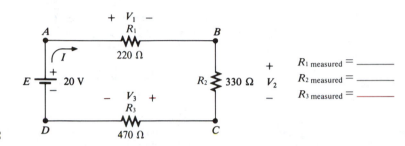

R_1 measured = _____

R_2 measured = _____

R_3 measured = _____

FIG. 4.2

Measure the voltage across the supply and across each resistor with the DMM (or VOM), making sure to note the polarity of the voltage across each element. Indicate the polarities on Fig. 4.3, and record the voltages below.

$E = $ _____ , $V_1 = $ _____ , $V_2 = $ _____ ,

$V_3 = $ _____

From your readings of the voltages, is Kirchhoff's voltage law satisfied around path $ABCD$? That is, does $E = V_1 + V_2 + V_3$?

(b) Now calculate the current through each resistor and record (use measured resistor values).

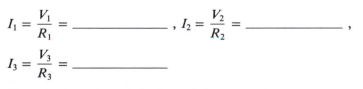

$$I_1 = \frac{V_1}{R_1} = \text{_____} \quad , \quad I_2 = \frac{V_2}{R_2} = \text{_____} \quad ,$$

$$I_3 = \frac{V_3}{R_3} = \text{_____}$$

How do the values of I_1, I_2, and I_3 compare?

What can you deduce about the characteristics of a series circuit from the current calculations above?

(c) The current I can be measured by inserting the multimeter (in the milliammeter mode) in series with the source as shown in Fig. 4.3.

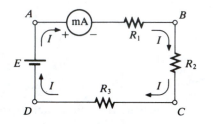

FIG. 4.3

Measure the current at positions A, B, C, and D of Fig. 4.3 and record below.

I (at A) = _____ , I (at B) = _____ , I (at C) = _____ , I (at D) = _____

Do your results verify the calculations of part 1(b)?

(d) Compute the total resistance of the circuit using the following equation:

$$R_T = \frac{E}{I}$$

$R_T =$ _____

(e) Using the measured values of the resistors from Table 4.2, calculate the total resistance R_T, where $R_T = R_1 + R_2 + R_3$. Record the value.

$R_T =$ _____

(f) Shut down and disconnect the power supply. Then measure the input resistance (R_T) across points A-D using the DMM (or VOM). Record that value.

R_T (measured) = _____

Compare this value with those in parts 1(d) and 1(e).

(g) Using the voltage divider rule, calculate the value of the voltage across resistor R_2, and then compare this value with the measured value from part 1(a).

V_2 (voltage divider rule) = _____ ,

V_2 [from part 1(a)] = _____

Part 2   Voltage Divider Rule

Construct the circuit of Fig. 4.4.

(a) Calculate V_1 using the voltage divider rule and measured resistance values.

$V_1 =$ _____

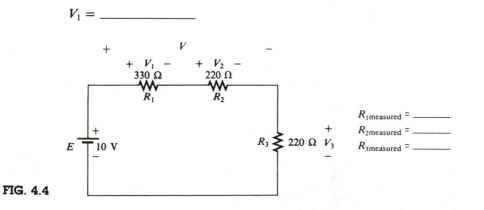

$R_{1measured} =$ _____

$R_{2measured} =$ _____

$R_{3measured} =$ _____

FIG. 4.4

(b) Measure V_1 and compare with the calculated value of part 2(a).

$V_1 = $ _____

(c) Measure V_2 and V_3 and comment on whether equal resistors in series have equal voltage drops across them.

$V_2 = $ _____ , $V_3 = $ _____

(d) Calculate the voltage V using the voltage divider rule.

$V = $ _____

(e) Measure the voltage V and compare with the value of part 2(d).

$V = $ _____

(f) Using measured resistor values, calculate the current I.

$I = $ _____

(g) Measure the current I and compare with the result of part 2(f).

$I = $ _____

PROBLEMS

1. Given the voltage V_1 in Fig. 4.5, determine the resistor R_1 using the voltage divider rule.

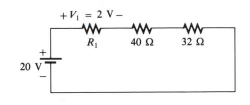

FIG. 4.5

$R_1 =$

2. Determine V_1 for the configuration of Fig. 4.6.

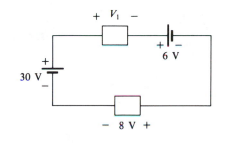

FIG. 4.6

$V_1 =$ _____

Parallel Resistance

OBJECT

To determine the total resistance of parallel dc circuits.

EQUIPMENT REQUIRED

Resistors

1—100-Ω, 1-kΩ, 1.2-kΩ, 2.2-kΩ, 1-MΩ

3—3.3-kΩ

Instruments

1—DMM (or VOM)

1—dc Power supply

EQUIPMENT ISSUED

TABLE 5.1

Item	Manufacturer and Model No.	Laboratory Serial No.
DMM (or VOM)		
Power supply		

TABLE 5.2

Resistors	
Nominal Value	Measured Value
100 Ω	
1 kΩ	
1.2 kΩ	
2.2 kΩ	
3.3 kΩ	
3.3 kΩ	
3.3 kΩ	
1 MΩ	

RÉSUMÉ OF THEORY

In a parallel circuit (Fig. 5.1), the total or equivalent resistance (R_T) is given by

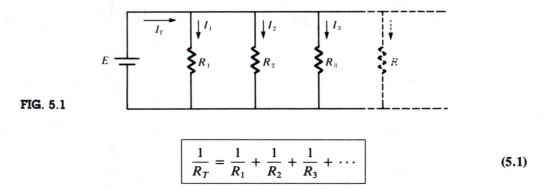

FIG. 5.1

$$\frac{1}{R_T} = \frac{1}{R_1} + \frac{1}{R_2} + \frac{1}{R_3} + \cdots$$

(5.1)

If there are only two resistors in parallel, it is more convenient to use

$$R_T = \frac{R_1 R_2}{R_1 + R_2}$$

(5.2)

For equal parallel resistors, the total resistance is equal to the resistance of one resistor divided by the number of resistors in parallel. That is,

$$R_T = \frac{R}{N} \tag{5.3}$$

For an unknown system the same technique applied for series dc circuits can be applied to parallel dc circuits. In Fig. 5.2 the total resistance can be determined from

$$R_T = \frac{E}{I} \tag{5.4}$$

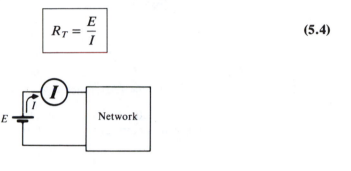

FIG. 5.2

For parallel resistors, the total resistance is always less than the value of the smallest parallel resistor. In some instances the total resistance is very close to the value of one resistor if that resistance is much less than the others.

PROCEDURE

Part 1   Two Parallel Resistors

(a) Construct the circuit of Fig. 5.3.

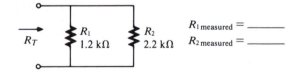

FIG. 5.3

Insert the measured values of the resistors in the spaces provided.

(b) Using the measured resistor values, calculate the total resistance using Eq. (5.1). Show all work.

$R_T =$ _____

(c) Using the ohmmeter section of your DMM or VOM, measure the total resistance of the network of Fig. 5.3.

$R_T =$ _____

(d) How do the results of parts 1(b) and 1(c) compare? Calculate their percent difference from

$$\% \text{ Difference} = \frac{|\text{Measured} - \text{Calculated}|}{\text{Calculated}} \times 100\% \qquad (5.5)$$

% Difference = _____

(e) Calculate the total resistance using measured values and Eq. (5.2). How do the results compare with part 1(b)?

$R_T =$ _____

(f) Apply 10 V to the circuit and insert an ammeter as shown in Fig. 5.4. Set the supply voltage with the voltmeter section of your meter and then insert the meter in the ammeter mode as shown. Record the ammeter reading below.

$I =$ _____

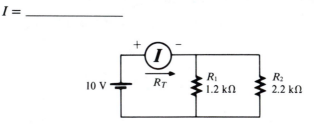

FIG. 5.4

(g) Using the supply voltage and the ammeter reading of part 1(f), calculate the total resistance using Eq. (5.4).

$R_T =$ _____

How do the results of this part and part 1(c) compare?

(h) How close is the total resistance to the resistance of the smaller resistor of Fig. 5.3? Is it less than R_1, and is it close to the magnitude of R_1?

Part 2   Three Parallel Resistors

(a) Construct the network of Fig. 5.5.

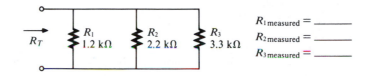

$R_{1\,measured} =$ _____

$R_{2\,measured} =$ _____

$R_{3\,measured} =$ _____

FIG. 5.5

Insert the measured values of the resistors in the spaces provided.

(b)   Calculate the total resistance using the measured resistor values.

$R_T =$ _____

(c) Measure the total resistance and calculate the percent difference.

$R_T =$ _____

% Difference = _____

(d) Apply 10 V and insert an ammeter as shown in Fig. 5.2, and record the source current below.

$I =$ _____

(e) Using the supply voltage and the ammeter reading of part 2(d), calculate the total resistance using Eq. (5.4).

$R_T = $ _____

How do the results of this part and part 2(c) compare?

(f) How close is the total resistance to the resistance of the smallest parallel resistor? Is it less than R_1, and is it close to the magnitude of R_1?

Has the addition of the 3.3-kΩ resistor made the total resistance closer to R_1 than it was in part 1(h)? Can you generate a conclusion based on this result?

Part 3　Equal Parallel Resistors

(a) Construct the network of Fig. 5.6.

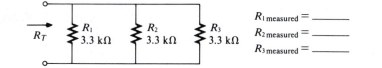

FIG. 5.6

R_1 measured $=$ _____
R_2 measured $=$ _____
R_3 measured $=$ _____

Insert the measured resistor values in the spaces provided.

(b) Using the nominal resistor values (all 3.3 kΩ), calculate the total resistance using Eq. (5.3).

$R_T = $ _____

(c) Measure the total resistance using the ohmmeter section of your DMM or VOM.

$R_T =$ _____

How do the calculated and measured values compare? Calculate the percent difference using Eq. (5.5).

% Difference = _____

(d) Calculate the total resistance using the measured values of each resistor.

$R_T =$ _____

How does this calculated value compare with the measured value of part 3(c)? What is the percent difference?

% Difference = _____

Part 4   A Composite

(a) Construct the network of Fig. 5.7.

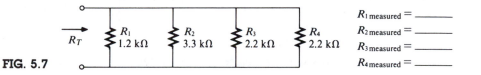

$R_{1\,measured} =$ _____
$R_{2\,measured} =$ _____
$R_{3\,measured} =$ _____
$R_{4\,measured} =$ _____

FIG. 5.7

Insert the measured resistor values in the spaces provided.

(b) Calculate the total resistance using the measured values.

$R_T =$ _____

(c) Measure the total resistance and find the percent difference between the calculated and measured values.

$R_T = $ _____

% Difference = _____

(d) Apply 10 V and measure the source current I. Then calculate the total resistance and compare with the measured value of part 4(c).

$I = $ _____

$R_T = $ _____

% Difference = _____

Part 5   Different Levels of Resistance

(a) Construct the network of Fig. 5.8.

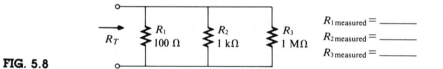

FIG. 5.8

$R_{1 \, measured} = $ _____

$R_{2 \, measured} = $ _____

$R_{3 \, measured} = $ _____

Insert the measured resistor values in the spaces provided.

(b) Using the measured values, calculate the total resistance of the circuit.

$R_T = $ _____

(c) Using an ohmmeter, measure the total resistance.

$R_T = $ _____

Calculate the percent difference between the measured and calculated values.

% Difference = _____

(d) Calculate the total resistance using measured values if the resistance R_3 is ignored.

$R_T =$ _____

What is the percent difference between this R_T and the measured value of part 5(c)?

% Difference = _____

(e) What is the total resistance using the measured values if resistances R_2 and R_3 are ignored.

$R_T =$ _____

What is the percent difference between this R_T and the measured value of part 5(c)?

% Difference = _____

(f) It was noted in the Résumé of Theory that the total resistance of parallel resistors is always less than the value of the smallest resistor. Was this true in parts 5(b) and 5(c)?

Using the results of parts 5(d) and 5(e), what general conclusion can you come to regarding the total resistance of parallel resistors when one resistor is considerably smaller (10:1 or greater) than the other parallel resistors?

Parallel dc Circuits

OBJECT

To investigate the characteristics of parallel dc circuits

EQUIPMENT REQUIRED

Resistors

$1-1.2$-kΩ, 3.3-kΩ

$2-2.2$-kΩ

Instruments

$1-$DMM (or VOM)

$1-$dc Power supply

EQUIPMENT ISSUED

TABLE 6.1

Item	Manufacturer and Model No.	Laboratory Serial No.
DMM (or VOM)		
Power supply		

TABLE 6.2

Resistors	
Nominal Value	Measured Value
1.2 kΩ	
2.2 kΩ	
2.2 kΩ	
3.3 kΩ	

RÉSUMÉ OF THEORY

In a parallel circuit (Fig. 6.1), the voltage across parallel elements is the same. Therefore,

$$V_1 = V_2 = V_3 = E \qquad\qquad (6.1)$$

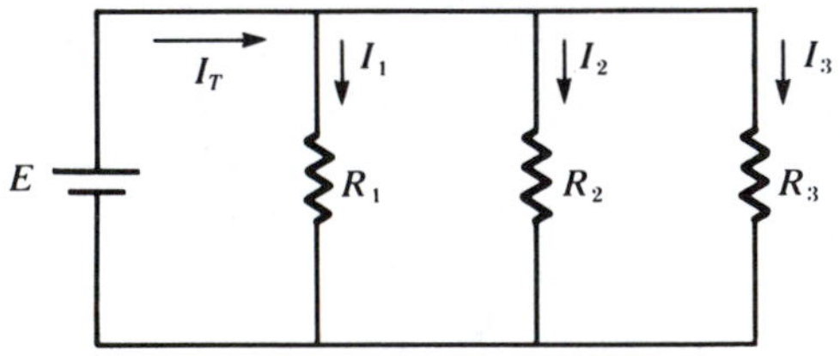

FIG. 6.1

For the network of Fig. 6.1, the currents are related by the following expression:

$$I_T = I_1 + I_2 + I_3 + \cdots \qquad\qquad (6.2)$$

and the current through each resistor is simply determined by Ohm's law:

$$I_1 = \frac{E}{R_1}, \ I_2 = \frac{E}{R_2}, \ I_3 = \frac{E}{R_3}, \ \text{etc.} \qquad\qquad (6.3)$$

The current divider rule (CDR) states that the current through one of two parallel branches is equal to the resistance of the *other* branch divided by the sum of the resistances of the two parallel branches and multiplied by the total current entering the two parallel branches. That is, for the network of Fig. 6.2,

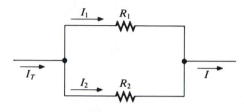

FIG. 6.2

$$I_1 = \frac{R_2 I_T}{R_1 + R_2} \quad \text{and} \quad I_2 = \frac{R_1 I_T}{R_1 + R_2}$$

(6.4)

PROCEDURE

Part 1   Unequal Parallel Resistors

(a) Construct the network of Fig. 6.3. Insert the measured values of the resistors in the spaces provided.

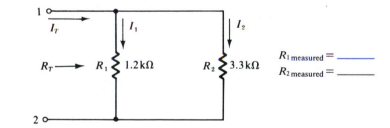

$R_{1 \text{ measured}} = \underline{\hspace{2cm}}$

$R_{2 \text{ measured}} = \underline{\hspace{2cm}}$

FIG. 6.3

(b) Using the measured values, calculate the total resistance of the network.

R_T (calculated) = \underline{\hspace{3cm}}

(c) Measure R_T with the ohmmeter section and compare with the result of part 1(b).

R_T(measured) = _____

(d) How does the total resistance compare with that of the smaller of the two parallel resistors? Will this always be true?

(e) Apply 20 V to points 1-2 of Fig. 6.3 and verify that the voltage across each resistor is 20 V.

V_{R_1} = _____ , V_{R_2} = _____

Calculate the currents through R_1 and R_2 and determine the total current I_T.

I_1 = _____ , I_2 = _____ , I_T = _____

(f) Measure I_T and compare with the result of part 1(e). Make sure the milliammeter is in series with the voltage source.

I_T(measured) = _____

Part 2 Equal Parallel Resistors

(a) Construct the network of Fig. 6.4. Insert the measured values of the resistors in the spaces provided.

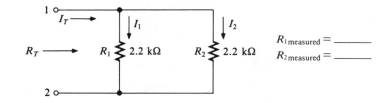

FIG. 6.4

(b) Calculate the total resistance assuming exactly equal resistor values. Use the nominal color-coded value.

$R_T =$ _____

(c) Connect the ohmmeter to points 1-2 and record the reading below. Compare with the result of part 2(b).

$R_{ohmmeter} =$ _____

(d) Using the measured resistance levels of R_1 and R_2, determine the total resistance.

$R_T =$ _____

(e) Compare the value obtained in part 2(d) with that obtained in part 2(c). Would you recommend using measured (rather than color-coded nominal) values in the future?

(f) Apply 10 V to points 1-2 of Fig. 6.4 and verify that the voltage across R_1 and R_2 is 10 V.

$V_{R_1} =$ _____ , $V_{R_2} =$ _____

(g) Calculate the currents I_1 and I_2 using Ohm's law, and determine I_T using Kirchhoff's current law.

$I_1 =$ _____ , $I_2 =$ _____ , $I_T =$ _____

(h) Measure the currents I_1, I_2, and I_T using the milliammeter range of your meter and compare with the results of part 2(g).

I_1 (measured) = _____ , I_2 (measured) = _____ ,

I_T (measured) = _____

Part 3   Three Parallel Resistors

(a) Construct the network of Fig. 6.5. Insert the measured value of each resistor.

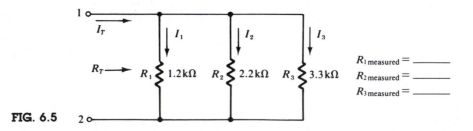

FIG. 6.5

(b) Using the measured values, calculate the total resistance R_T.

R_T(calculated) = _____

(c) Determine R_T using the ohmmeter section of the multimeter and compare with the result of part 3(b).

R_T(measured) = _____

(d) Calculate the currents I_1, I_2, and I_3 if a source voltage of 10 V is applied. Calculate the total input current I_T using Kirchhoff's current law.

I_1 = _____ , I_2 = _____ , I_3 = _____

I_T = _____

(e) Apply 10 V to the network of Fig. 6.5 and measure the currents I_3 and I_T. Compare with the results of part 3(d).

I_3 = _____ , I_T = _____

(f) Is the total resistance less than the magnitude of the smallest resistance? Comment accordingly.

Rheostats and Potentiometers

OBJECT

To study the characteristics and applications of a linear rheostat and potentiometer.

EQUIPMENT REQUIRED

Resistors

1—1-kΩ

1—0–1-kΩ linear carbon potentiometer

Instruments

1—DMM (or VOM)

1—dc Power supply

EQUIPMENT ISSUED

TABLE 7.1

Item	Manufacturer and Model No.	Laboratory Serial No.
DMM (or VOM)		
Power supply		

TABLE 7.2

Resistors	
Nominal Value	Measured Value
1 kΩ	

RÉSUMÉ OF THEORY

The rheostat is a two-terminal variable resistance device whose terminal characteristics can be linear (straight-line) or nonlinear. It is typically used in series with the load to control the voltage or current levels.

The potentiometer (often referred to as a *pot*) is a three-terminal device that is used primarily to control *potential* (voltage) levels. The potentiometer can reduce the voltage across a load to zero, while a two-terminal rheostat can reduce it only to a level determined by its maximum resistance and the resistance of the load.

A three-terminal potentiometer can be used as a two-terminal rheostat if desired. We will examine both possibilities in this experiment.

PROCEDURE

Part 1 Potentiometer Characteristics

(a) The graphic symbol for the potentiometer and the actual device appear in Fig. 7.1. As indicated by the schematic representation, the resistance between the outside terminals is fixed at the full value while the resistance between the movable contact and either outside terminal is a function of the position of the "wiper arm."

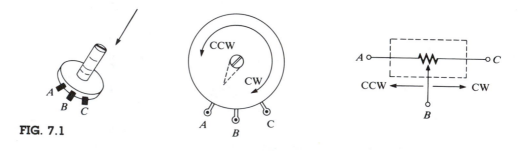

FIG. 7.1

(b) Turn the shaft to the full CW position and mark the position of the shaft on Fig. 7.2(b) as shown in Fig. 7.2(a).

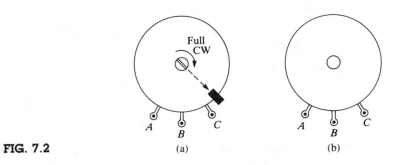

FIG. 7.2   (a)   (b)

(c) Turn the shaft to the full CCW position and mark the position of the shaft on Fig. 7.2(b) as shown in Fig. 7.3.

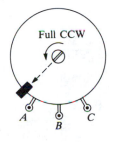

FIG. 7.3

(d) Break down the region between full CW and full CCW of the potentiometer of Fig. 7.2(b) as demonstrated by Fig. 7.4. In other words, mark the locations of a 1/4, a 1/2, and a 3/4 turn from the full CCW position to the CW position. We will need these positions for the measurements to be performed.

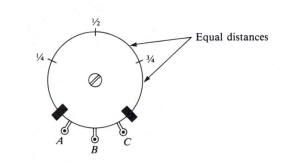

FIG. 7.4

(e) Connect the DMM (or VOM) to the terminals indicated in Table 7.3 and record the resistance value for each setting.

TABLE 7.3

Potentiometer Setting	Resistance To Be Measured		
	R_{AC}	R_{AB}	R_{BC}
Fully CCW (A)			
1/4 turn CW from A			
1/2 turn CW from A			
3/4 turn CW from A			
Fully CW (C)			

(f) Plot the resistances R_{AC}, R_{AB}, and R_{BC} versus position on Graph 7.1.

(g) Determine R_{AC} from $R_{AC} = R_{AB} + R_{BC}$ at the 1/4 turn position.

$R_{AC} = R_{AB} + R_{BC} =$ _____ + _____ = _____

Repeat the above for the 1/2 turn position.

$R_{AC} =$ _____

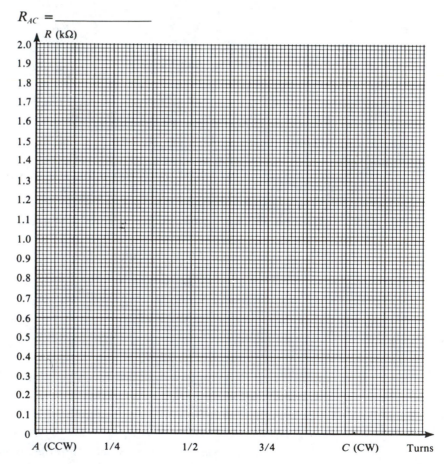

GRAPH 7.1

How do the values obtained above for R_{AC} compare with the measured values?

(h) Does R_{AC} (the resistance between the two outermost terminals) remain fairly constant throughout the measurement span? Why?

(i) Is the magnitude (ignore the sign) of the slope of R_{AB} the same as for R_{BC}? Should it be? Why?

(j) At 1/2 turn, does $R_{AB} = R_{BC}$? Should it be? Why?

Part 2 Potentiometer Control of Potential (Voltage) Levels

(a) Construct the network of Fig. 7.5.
Set the dc supply voltage with the DMM (or VOM).

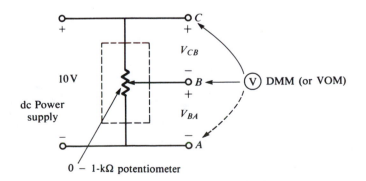

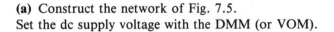

FIG. 7.5 $0 - 1$-kΩ potentiometer

Using the VOM and DMM, read the voltages V_{CA}, V_{BA}, and V_{CB} at each step of the sequence indicated in Table 7.4. Be sure to read both voltages before changing the setting. Do not change the setting based on your readings. Leave the setting as determined by the judgments made on Fig. 7.2(b).

TABLE 7.4

Potentiometer Setting	V_{CA}	V_{BA}	V_{CB}
Fully CCW (A)			
1/4 turn CW from A			
1/2 turn CW from A			
3/4 turn CW from A			
Fully CW (C)			

(b) Plot the voltages V_{CA}, V_{BA}, and V_{CB} versus position on Graph 7.2.

(c) Which voltage remains independent of the position of the rotor? Why?

(d) What are the minimum and maximum values of V_{BA} and V_{CB}?

Why do they have the above maximum and minimum values?

(e) How would you compare the magnitudes of the slopes of the curves for V_{BA} and V_{CB}?

(f) At the 5/8 turn position on Graph 7.2, does $V_{CA} = V_{CB} + V_{BA}$ using the graphical data?

$V_{CB} = $ _____

$V_{BA} = $ _____

$V_{CA} = $ _____

If it does, explain why.

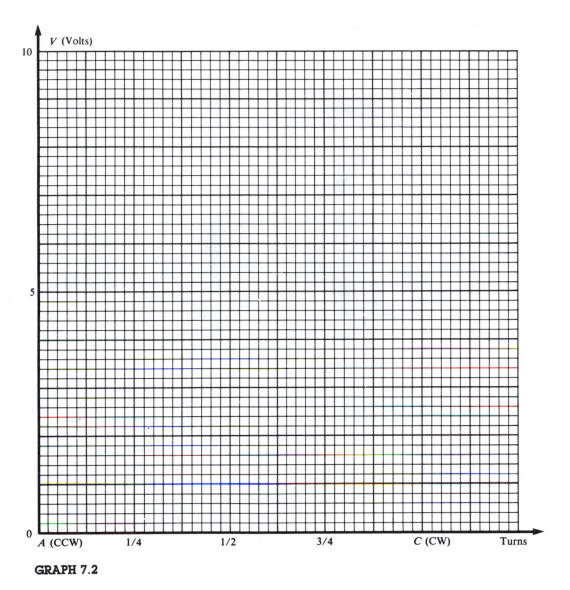

GRAPH 7.2

Part 3   Voltage Control

        (a) Construct the circuit of Fig. 7.6.

        Read the voltage V_{CB} at each step indicated in Table 7.5. Calculate the current I using the measured resistance value.

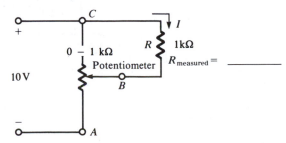

FIG. 7.6

TABLE 7.5

Potentiometer Setting	V_{CB}	$I = \dfrac{V_{CB}}{R_{\text{measured}}}$
Fully CCW (A)		
1/4 turn CW from A		
1/2 turn CW from A		
3/4 turn CW from A		
Fully CW (C)		

(b) Plot the voltage V_{CB} versus position on Graph 7.3.

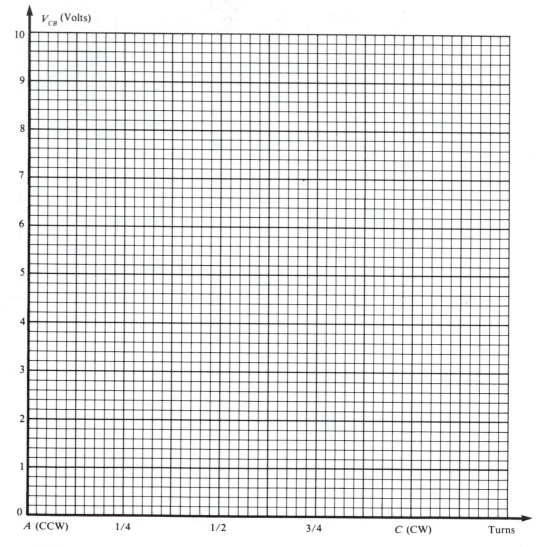

GRAPH 7.3

(c) Is the curve linear or nonlinear? Why?

(d) Plot the curve of the current I versus position on Graph 7.4.

(e) Does the potentiometer also offer a form of current control? Is this relationship linear or nonlinear?

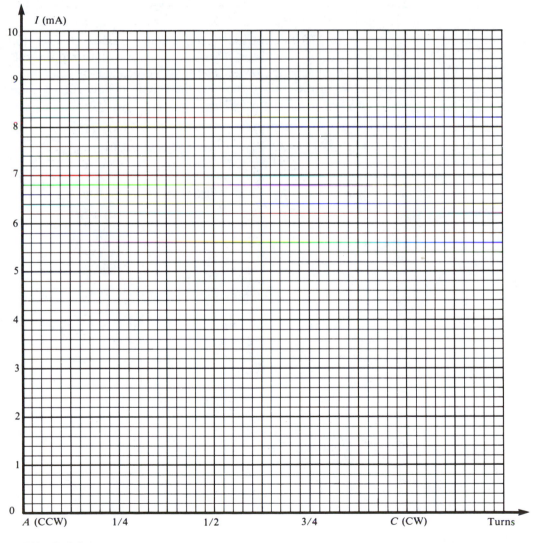

GRAPH 7.4

Part 4 Current Control with a Two-Point Rheostat

(a) Construct the network of Fig. 7.7.

Note that only two terminals of the potentiometer are being used. The total resistance of the circuit is now $R_{BC} + R$.

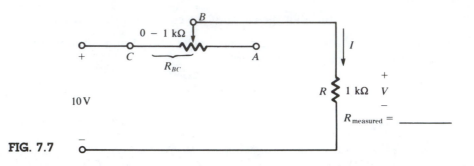

FIG. 7.7

(b) Set the rheostat to the minimum resistance level (fully CW) to establish 10 V across the 1-kΩ resistor. Then adjust the rheostat until the voltage across the 1-kΩ resistor is 9 V [using the DMM (or VOM)]. Calculate the resistance R_{BC} (using the measured values of the 1-kΩ resistor and the voltage divider rule) required to set this voltage level, and insert in Table 7.6. That is,

$$V = \frac{RE}{R + R_{BC}} = \frac{(1000)(10)}{1000 + R_{BC}}$$

or

$$(1000 + R_{BC})V = 10,000$$

$$1000 + R_{BC} = \frac{10,000}{V}$$

and

$$R_{BC} = \frac{10,000}{V} - 1000$$

Disconnect the power supply at point C and measure the resistance R_{BC} with the ohmmeter section of the meter and insert in Table 7.6. Repeat the above for the other voltage levels indicated in Table 7.6 and calculate the current I.

TABLE 7.6

V	R_{BC} (calculated)	R_{BC} (measured)	$I = \dfrac{V_R}{R_{\text{measured}}}$
10 V	0 Ω	0 Ω	10 mA
9 V			
8 V			
7 V			
6 V			
5 V			

(c) What is the minimum level the voltage V can be set to? Why is it limited to this level?

(d) Using the data obtained in Table 7.6, comment on the rheostat as a current control device.

(e) Are you satisfied with the way the calculated and measured values of R_{BC} compare? Comment accordingly.

Series-Parallel
dc Circuits

OBJECT

To investigate the characteristics of series-parallel dc circuits.

EQUIPMENT REQUIRED

Resistors

1—91-Ω, 220-Ω, 330-Ω, 470-Ω

Instruments

1—DMM (or VOM)
1—dc Power supply

EQUIPMENT ISSUED

TABLE 8.1

Item	Manufacturer and Model No.	Laboratory Serial No.
DMM (or VOM)		
Power supply		

TABLE 8.2

Resistors	
Nominal Value	**Measured Value**
91 Ω	
220 Ω	
330 Ω	
470 Ω	

RÉSUMÉ OF THEORY

The analysis of series-parallel dc networks requires a firm understanding of the basics of both series and parallel networks. In the series-parallel configuration, you will have to isolate series and parallel configurations and make the necessary combinations for reduction as you work toward the desired unknown quantity.

  As a rule, it is best to make a mental sketch of the path you plan to take toward the complete solution before introducing the numerical values. This may result in savings in both time and energy. Always work with the isolated series or parallel combinations in a branch before tying the branches together. For complex networks, a carefully redrawn set of reduced networks may be required to ensure that the unknowns are conserved and that every element has been properly included.

PROCEDURE

Part 1

  (a) Construct the series-parallel network of Fig. 8.1, and insert the measured resistor values in the spaces provided.

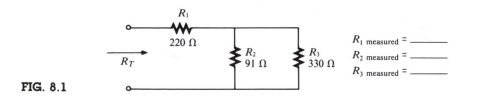

FIG. 8.1

(b) Calculate the total resistance R_T using the measured resistance values.

R_T(calculated) = _____

(c) Use the ohmmeter section of the multimeter to measure R_T and compare with the result of part 1(b).

R_T(measured) = _____

(d) If 10 V were applied as shown in Fig. 8.2, *calculate* the currents I_T, I_1, I_2, and I_3 using the measured resistor values.

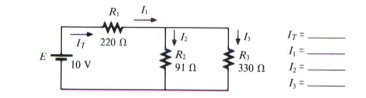

FIG. 8.2

(e) Apply 10 V and *measure* the currents I_1, I_2, and I_3 using the milliammeter section of your multimeter. Compare with the results of part 1(d). Be sure the meter is in series with the resistor through which the current is to be measured.

I_1(measured) = _____ , I_2(measured) = _____ ,

I_3(measured) = _____

(f) Using the results of part 1(d), *calculate* the voltages V_1, V_2, and V_3.

$V_1 =$ _____ , $V_2 =$ _____ , $V_3 =$ _____

(g) *Measure* the voltages V_1, V_2, and V_3 and compare with the results of part 1(f).

$V_1 =$ _____ , $V_2 =$ _____ , $V_3 =$ _____

(h) Referring to Fig. 8.2, does $E = V_1 + V_2$? Use the measured values of part 1(g) to check on this equality.

Part 2

(a) Construct the series-parallel network of Fig. 8.3.

(b) Calculate the total resistance R_T using the measured value of each resistor.

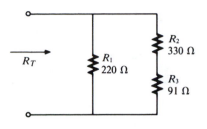

FIG. 8.3

R_T(calculated) = _____

(c) Use the ohmmeter section of your multimeter to measure the total resistance R_T. Compare with the result of part 2(b).

R_T(measured) = _____

(d) If 10 V were applied to the network as shown in Fig. 8.4, *calculate* the currents I_T, I_1, I_2, and I_3.

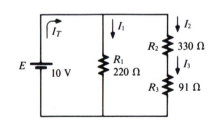

FIG. 8.4

$I_T =$ _____ , $I_1 =$ _____ , $I_2 =$ _____

$I_3 =$ _____

(e) Apply 10 V and measure the currents I_T, I_1, I_2, and I_3. Compare with the results of part 2(d).

I_T(measured) = _____ , I_1(measured) = _____ ,

I_2(measured) = _____ , I_3(measured) = _____

(f) Using the results of part 2(d), calculate the voltages V_1, V_2, and V_3.

V_1 = _____ , V_2 = _____ , V_3 = _____

(g) Measure the voltages V_1, V_2, and V_3 and compare with the results of part 2(f).

V_1 = _____ , V_2 = _____ , V_3 = _____

(h) Does $E = V_2 + V_3$? Should it? Verify your statement using the data of part 2(g).

Part 3

(a) Construct the series-parallel network of Fig. 8.5 and insert the measured value of each resistor.

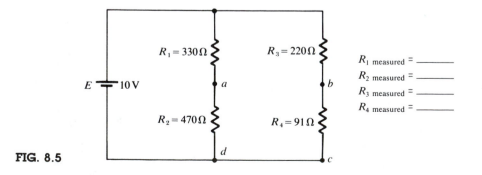

FIG. 8.5

(b) Using the measured resistor values, calculate the voltages V_{ad} and V_{bc}.

$V_{ad} = $ _____ , $V_{bc} = $ _____

(c) Measure the voltages V_{ad} and V_{bc} and compare with the results of part 3(b).

$V_{ad} = $ _____ , $V_{bc} = $ _____

(d) Using the results of part 3(b), calculate the voltage V_{ab}.

V_{ab} (calculated) = _____

(e) Measure the voltage V_{ab} and compare with the result of part 3(d).

V_{ab} (measured) = _____

Part 4

(a) Construct the network of Fig. 8.6. Insert the measured value of each resistor.

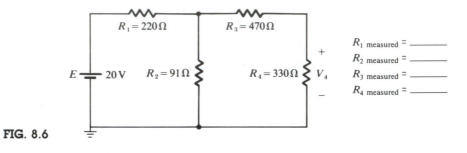

FIG. 8.6

(b) Calculate the voltage V_4 using the measured resistor values.

V_4 (calculated) = _____

(c) Measure the voltage V_4 and compare with the result of part 4(b).

V_4 (measured) = _____

Superposition Principle (dc)

OBJECT

*To verify experimentally the superposition
principle as applied to dc circuits.*

EQUIPMENT REQUIRED

Resistors

1—1.2-kΩ, 2.2-kΩ, 3.3-kΩ

Instruments

1—DMM (or VOM)

2*—dc Power supplies

*The unavailability of two power supplies may require that two groups work
together.

EQUIPMENT ISSUED

TABLE 9.1

Item	Manufacturer and Model No.	Laboratory Serial No.
DMM (or VOM)		
Power supply		
Power supply		

TABLE 9.2

Resistors	
Nominal Value	**Measured Value**
1.2 kΩ	
2.2 kΩ	
3.3 kΩ	

RÉSUMÉ OF THEORY

The superposition principle states that the current through, or voltage across, any resistive branch of a multisource network is the algebraic sum of the contributions due to each source acting independently. When the effects of one source are considered, the others are replaced by their internal resistances. Superposition is effective only for linear circuit relationships.

This principle permits one to analyze circuits without resorting to simultaneous equations. Nonlinear effects, such as power, which varies as the square of the current or voltage, cannot be analyzed using this principle.

PROCEDURE

Part 1   Calculated Results

Insert the measured resistor values in Fig. 9.1.

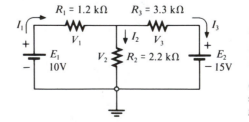

R_1 measured = _____
R_2 measured = _____
R_3 measured = _____

Caution: Be sure dc supplies have common ground.

FIG. 9.1

(a) Calculate the currents I_1, I_2, and I_3 for the source E_1 using the diagram of Fig. 9.2 and the measured resistor values.

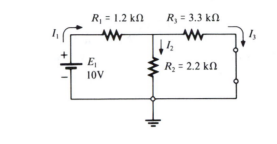

FIG. 9.2

$I_1 =$ _____

$I_2 =$ _____

$I_3 =$ _____

(b) Calculate the currents I_1, I_2, and I_3 for the source E_2 using the diagram of Fig. 9.3 and the measured resistor values.

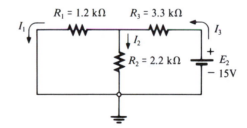

FIG. 9.3

$I_1 =$ _____

$I_2 =$ _____

$I_3 =$ _____

(c) Determine I_1, I_2, and I_3 for the network of Fig. 9.1 using the calculated results of parts 1(a) and 1(b). Show their resulting directions on Fig. 9.1.

$I_1 =$ _____

$I_2 =$ _____

$I_3 =$ _____

(d) Using the results of part 1(a), calculate the voltage across each resistor using Ohm's Law. Insert the polarities of V_1, V_2, and V_3 on Fig. 9.2.

$V_1 =$ _____

$V_2 =$ _____

$V_3 =$ _____

(e) Using the results of part 1(b), calculate the voltage across each resistor using Ohm's Law. Insert the polarities of V_1, V_2, and V_3 on Fig. 9.3.

$V_1 =$ _____

$V_2 =$ _____

$V_3 =$ _____

(f) Determine V_1, V_2, and V_3 for the network of Fig. 9.1 using the calculated results of parts 1(d) and 1(e). Insert their polarities on Fig. 9.1.

(g) Using the results of part 1(a), calculate the power to each resistor.

$P_1 = $ _____

$P_2 = $ _____

$P_3 = $ _____

(h) Using the results of part 1(b), calculate the power to each resistor.

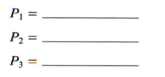

$P_1 = $ _____

$P_2 = $ _____

$P_3 = $ _____

(i) Using the results of part 1(c), calculate the power to each resistor.

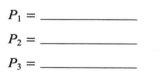

$P_1 = $ _____

$P_2 = $ _____

$P_3 = $ _____

(j) Add the power levels determined in parts 1(g) and 1(h) and compare with the levels determined in part 1(i).

$P_1 [1(g)] + P_1 [1(h)] = $ _____

$P_1 [1(i)] = $ _____

$P_2 [1(g)] + P_2 [1(h)] = $ _____

$P_2 [1(i)] = $ _____

$$P_3 \ [1(\text{g})] + P_3 \ [1(\text{h})] = \underline{\hspace{2cm}}$$

$$P_3 \ [1(\text{i})] = \underline{\hspace{2cm}}$$

How do the results compare? Why?

Part 2 Measured Values

(a) Construct the network of Fig. 9.2 and measure the voltages V_1, V_2, and V_3.

$V_1 = \underline{\hspace{2.5cm}}$

$V_2 = \underline{\hspace{2.5cm}}$

$V_3 = \underline{\hspace{2.5cm}}$

How do they compare with the results of part 1(d)?

(b) Construct the network of Fig. 9.3 and measure the voltages V_1, V_2, and V_3.

$V_1 = \underline{\hspace{2.5cm}}$

$V_2 = \underline{\hspace{2.5cm}}$

$V_3 = \underline{\hspace{2.5cm}}$

How do they compare with the results of part 1(e)?

(c) Construct the network of Fig. 9.1 and measure the voltages V_1, V_2, and V_3.

$V_1 = \underline{\hspace{2.5cm}}$

$V_2 = \underline{\hspace{2.5cm}}$

$V_3 = \underline{\hspace{2.5cm}}$

How do they compare with the results of part 1(f)?

(d) Calculate the currents I_1, I_2, and I_3 from the measured values of part 2(a).

$I_1 =$ _____

$I_2 =$ _____

$I_3 =$ _____

How do they compare with the results of part 1(a)?

(e) Calculate the currents I_1, I_2, and I_3 from the measured values of part 2(b).

$I_1 =$ _____

$I_2 =$ _____

$I_3 =$ _____

How do they compare with the results of part 1(b)?

(f) Calculate the currents I_1, I_2, and I_3 from the results of part 2(c).

$I_1 =$ _____

$I_2 =$ _____

$I_3 =$ _____

How do they compare with the results of part 1(c)?

(g) Using the results of part 2(f), calculate the power to each resistor.

$P_1 =$ _____

$P_2 =$ _____

$P_3 =$ _____

How do they compare with the results of part 1(i)?

QUESTION

Has the superposition theorem been substantiated by the results of this experiment? Comment accordingly.

Thevenin's Theorem and Maximum Power Transfer

OBJECT

To verify Thevenin's theorem and the maximum power transfer principle.

EQUIPMENT REQUIRED

Resistors

1 — 91-Ω, 220-Ω, 330-Ω, 470-Ω

1 — 0–1-kΩ potentiometer

Instruments

1 — DMM (or VOM)

1 — dc Power supply

EQUIPMENT ISSUED

TABLE 10.1

Item	Manufacturer and Model No.	Laboratory Serial No.
DMM (or VOM)		
Power supply		

TABLE 10.2

Resistors	
Nominal Value	**Measured Value**
91 Ω	
220 Ω	
330 Ω	
470 Ω	

RÉSUMÉ OF THEORY

Through the use of Thevenin's theorem, a complex two-terminal, linear, multisource dc network can be replaced by one having a single source and resistor.

The Thevenin equivalent circuit consists of a single dc source referred to as the *Thevenin voltage* and a single fixed resistor called the *Thevenin resistance*. The Thevenin voltage is the open-circuit voltage across the terminals in question (Fig. 10.1). The Thevenin resistance is the resistance between these terminals with all of the voltage and current sources replaced by their internal resistances (Fig. 10.1).

If a dc voltage source is to deliver maximum power to a resistor, the resistor must have a value equal to the internal resistance of the source. In a complex network, maximum power transfer to a load will occur when the load resistance is equal to the

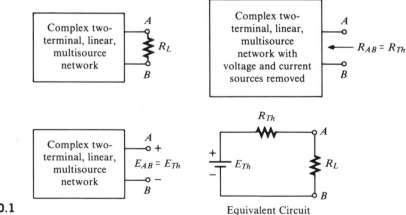

FIG. 10.1   Equivalent Circuit

Thevenin resistance "seen" by the load. For this value, the voltage across the load will be one-half of the Thevenin voltage. In equation form, the maximum power is given by

$$P_{\text{max}} = \frac{E_{Th}^2}{4R_{Th}}$$

PROCEDURE

Part 1 Thevenin's Theorem

Calculations:

(a) Insert the measured resistor values into Fig. 10.2 and *calculate* the Thevenin voltage and resistance for the network to the left of points *a-a'*.

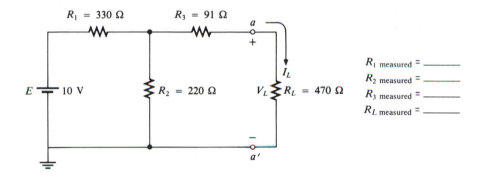

FIG. 10.2

$E_{Th} =$ _____ , $R_{Th} =$ _____ . Enter these values in column 1 of Table 10.3.

(b) Insert the values of E_{Th} and R_{Th} in Fig. 10.3 and calculate I_L.

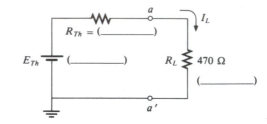

FIG. 10.3

        I_L = _____ . Enter this value in the first space in column 3 of Table 10.3.

(c) Calculate the current I_L in the original network of Fig. 10.2 using series-parallel techniques (use measured resistor values).

        I_L = _____ . Enter this value in the second space in column 3 of Table 10.3.

Measurements:

(d) Construct the network of Fig. 10.2 and measure the voltage V_L. Calculate the current I_L. Enter this value in the first space in column 4 of Table 10.3.

        V_L(measured) = _____ , I_L(calculated from V_L) = _____

R_{Th}:

(e) Determine R_{Th} by constructing the network of Fig. 10.4 and measuring the resistance between points a-a'.

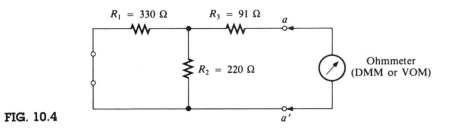

FIG. 10.4

$R_{Th} =$ _____ . Enter this value in the second space in column 2 of Table 10.3.

E_{Th}:

(f) Determine E_{Th} by constructing the network of Fig. 10.5 and measuring the open-circuit voltage between points a-a'.

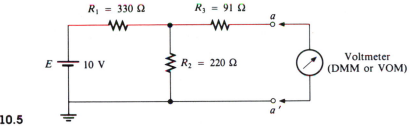

FIG. 10.5

E_{Th}(measured) = _____ . Enter this value in the first space in column 2 of Table 10.3.

Thevenin Network:

(g) Construct the network of Fig. 10.6 and set the values obtained for E_{Th} and R_{Th} in parts 1(e) and 1(f), respectively. Use the ohmmeter section of your meter to set the potentiometer properly. Then measure the voltage V_L and calculate the current I_L. Enter this value in the second space in column 4 of Table 10.3.

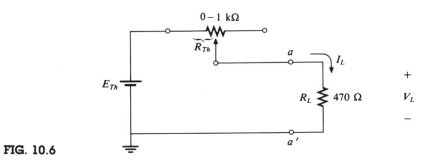

FIG. 10.6

V_L (measured) = _____ , I_L (calculated from V_1) = _____

TABLE 10.3

Calculated Values of E_{Th} and R_{Th} [Part 1(a)]	Measured Values of E_{Th} and R_{Th} [Parts 1(e) and 1(f)]	Calculated Values of I_L [Parts 1(b) and 1(c)]	Measured Values of I_L [Parts 1(d) and 1(g)]
E_{Th} = _____	E_{Th} = _____ [part 1(f)]	Using Thevenin equivalent: I_L = _____ [part 1(b)]	From Fig. 10.2: I_L = _____ [part 1(d)]
R_{Th} = _____	R_{Th} = _____ [part 1(e)]	Using other methods: I_L = _____ [part 1(c)]	Using Thevenin equivalent: I_L = _____ [part 1(g)]

How do the calculated and measured values of E_{Th} and R_{Th} compare?

How do the calculated and measured values of I_L compare?

Part 2   Maximum Power Transfer

(a) Construct the network of Fig. 10.7 and set the potentiometer to 50 Ω. Measure the voltage across R as you vary R through the following values: 50, 100, 200, 300, 330, 400, 600, 800, and 1000 Ω. Be sure to set the resistance with the ohmmeter section of

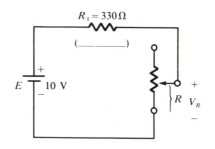

FIG. 10.7

your meter before each reading. (Remember to disconnect the dc supply when setting the resistance level.) Complete Table 10.4 and plot P_R (power) versus R on Graph 10.1.

   (b) Theoretically, for what value of R will the power delivered to R be maximum? Check this value against that obtained from the graph.

$R =$ _____

   In addition, what should the voltage across R be when R is set for maximum power? Do your experimental data substantiate this?

$V_R =$ _____

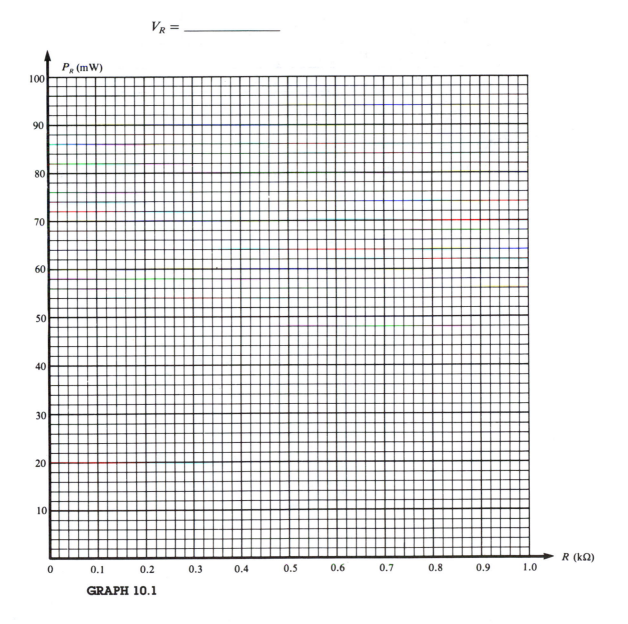

GRAPH 10.1

TABLE 10.4

R	V_R	$P_R = \dfrac{V_R^2}{R}$
50 Ω		
100 Ω		
200 Ω		
300 Ω		
R_1 (measured)		
400 Ω		
600 Ω		
800 Ω		
1000 Ω		

QUESTION

Is Thevenin's Theorem verified by the results of this experiment? Be specific.

Methods of Analysis

OBJECT

To become familiar with the branch-, mesh-, and nodal-analysis techniques.

EQUIPMENT REQUIRED

Resistors

1 — 1.0-kΩ, 1.2-kΩ

1 — 2.2-kΩ, 3.3-kΩ

Instruments

1 — DMM (or VOM)

2* — dc Power supplies

*The unavailability of two supplies will simply require that two groups work together.

EQUIPMENT ISSUED

TABLE 11.1

Item	Manufacturer and Model No.	Laboratory Serial No.
DMM (or VOM)		
Power supply		
Power supply		

TABLE 11.2

Resistors	
Nominal Value	Measured Value
1.0 kΩ	
1.2 kΩ	
2.2 kΩ	
3.3 kΩ	

RÉSUMÉ OF THEORY

The branch-, mesh-, and nodal-analysis techniques are used to solve complex networks with a single source or networks with more than one source that are not in series or parallel.

The branch- and mesh-analysis techniques will determine the currents of the network, while the nodal-analysis approach will provide the potential levels of the nodes of the network with respect to some reference.

The application of each technique follows a sequence of steps, each of which will result in a set of equations with the desired unknowns. It is then only a matter of solving these equations for the various variables, whether they be current or voltage.

PROCEDURE

Part 1   Branch-current Analysis

(a) Construct the network of Fig. 11.1 and insert the measured values of the resistors in the spaces provided.

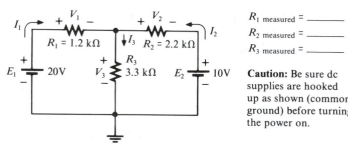

R_1 measured = _____
R_2 measured = _____
R_3 measured = _____

Caution: Be sure dc supplies are hooked up as shown (common ground) before turning the power on.

FIG. 11.1

(b) Using branch-current analysis, *calculate* the current through each branch of the network of Fig. 11.1 and insert in Table 11.3. Use the measured resistor values and assume the current directions shown in the figure.

TABLE 11.3

Current	Calculated	Measured
I_1		
I_2		
I_3		

(c) Measure the voltages V_1, V_2, and V_3 and enter below with a minus sign for any polarity that is opposite to that in Fig. 11.1.

$V_1 =$ _____ , $V_2 =$ _____ , $V_3 =$ _____

Calculate the currents I_1, I_2, and I_3 and insert in Table 11.3 as the measured values. Be sure to include a minus sign if the current direction is opposite to that appearing in Fig. 11.1.

How do the calculated and measured results compare?

Part 2   Mesh Analysis

(a) Construct the network of Fig. 11.2 and insert the measured values of the resistors in the spaces provided.

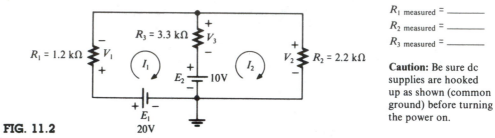

R_1 measured = _____

R_2 measured = _____

R_3 measured = _____

Caution: Be sure dc supplies are hooked up as shown (common ground) before turning the power on.

FIG. 11.2

(b) Using mesh analysis, *calculate* the mesh currents of the network. Use the measured resistor values and the indicated directions for the mesh currents. Then determine the current through each resistor and insert in Table 11.4 in the "Calculated" column.

Mesh currents: $I_1 =$ _____ , $I_2 =$ _____

TABLE 11.4

Current	Calculated	Measured
I_{R1}		
I_{R2}		
I_{R3}		

(c) Measure the voltages V_1, V_2, and V_3 and enter below with a minus sign for any polarity that is opposite to that in Fig. 11.2.

$V_1 = $ _____ , $V_2 = $ _____ , $V_3 = $ _____

Calculate the currents I_{R1}, I_{R2}, and I_{R3} and insert in Table 11.4 as the measured values. Be sure to include a minus sign if the current direction is opposite to that defined by the polarity of the voltage across the resistor. $I_{R1} = I_1$, $I_{R2} = I_2$, and $I_{R3} = I_1 - I_2$, as defined by Fig. 11.2.

How do the calculated and measured results compare?

Part 3   Nodal Analysis

(a) Construct the network of Fig. 11.3 and insert the measured resistor values.

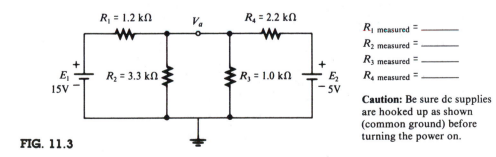

$R_1 = 1.2\ k\Omega$ V_a $R_4 = 2.2\ k\Omega$

E_1 15V $R_2 = 3.3\ k\Omega$ $R_3 = 1.0\ k\Omega$ E_2 5V

R_1 measured = _____
R_2 measured = _____
R_3 measured = _____
R_4 measured = _____

Caution: Be sure dc supplies are hooked up as shown (common ground) before turning the power on.

FIG. 11.3

Calculations:

    (b) Convert both voltage sources to current sources.

    (c) Determine V_a using nodal analysis.

V_a(calculated) = _____

    (d) Using V_a, calculate the values of I_{R1}, I_{R2}, I_{R3}, and I_{R4} and insert in Table 11.5.

TABLE 11.5

Current	Calculated	Measured
I_{R1}		
I_{R2}		
I_{R3}		
I_{R4}		

Measurements:

    (e) Energize the network and measure the voltage V_a. Compare with the result of part 3(c).

V_a(measured) = _____

(f) Calculate the currents I_{R1}, I_{R2}, I_{R3}, and I_{R4} and insert in Table 11.5 as the measured results.

(g) How do the calculated and measured results compare?

QUESTION

Many times one is faced with the question of which method to use in a particular problem. The laboratory activity does not prepare one to make such choices but only shows that the methods work and are solid. From your experience in this activity, summarize in your own words which method you would choose and why.

Capacitors

OBJECT

To become familiar with the characteristics of a capacitor in a dc system.

EQUIPMENT REQUIRED

Resistors

2—1.2-kΩ, 100-kΩ

Capacitors

2—100-μF (electrolytic)

Instruments

1—DMM (or VOM)

1—dc Power supply

1—Single-pole, single-throw switch

EQUIPMENT ISSUED

TABLE 12.1

Item	Manufacturer and Model No.	Laboratory Serial No.
DMM (or VOM)		
Power supply		

TABLE 12.2

Resistors	
Nominal Value	**Measured Value**
1.2 kΩ	
1.2 kΩ	
100 kΩ	
100 kΩ	

RÉSUMÉ OF THEORY

The resistor dissipates electrical energy in the form of heat. In contrast, the capacitor is a component that stores electrical energy in the form of an electrical field. The capacitance of a capacitor is a function of its geometry and the dielectric used. Dielectric materials (insulators) are rated by their ability to support an electric field in terms of a figure called the dielectric constant (k). Dry air is the standard dielectric for purposes of reference and is assigned the value of unity. Mica, with a dielectric constant of 6, has six times the capacitance of a similarly constructed air capacitor.

One of the most common capacitors is the parallel-plate type. The capacitance (in farads) is determined by

$$C = 8.85 \times 10^{-12} \epsilon_r \frac{A}{d} \qquad (12.1)$$

where ϵ_r is the relative permittivity (or dielectric constant), A is the area of the plates (m^2), and d is the distance between the plates (m). By changing any one of the three parameters, one can easily change the capacitance.

The dielectric constant is not to be confused with the dielectric strength of a material, which is given in volts per unit length and is a measure of the maximum stress that a dielectric can withstand before it breaks down and loses its insulator characteristics.

In a dc circuit, the volt-ampere characteristics of a capacitor in the steady-state mode are such that the capacitor prevents the flow of dc current but will charge up to a dc voltage. Essentially, therefore, the characteristics of a capacitor in the steady-state mode are those of an *open circuit*. The charge Q stored by a capacitor is given by

$$\boxed{Q = CV} \qquad\qquad\qquad \textbf{(12.2)}$$

where C is the capacitance and V is the voltage impressed across the capacitor.
Capacitors in series behave like resistors in parallel:

$$\boxed{\frac{1}{C_T} = \frac{1}{C_1} + \frac{1}{C_2} + \frac{1}{C_3} + \cdots + \frac{1}{C_N}} \qquad\qquad \textbf{(12.3)}$$

Capacitors in parallel behave like resistors in series:

$$\boxed{C_T = C_1 + C_2 + C_3 + \cdots + C_N} \qquad\qquad \textbf{(12.4)}$$

The energy stored by a capacitor (in joules) is determined by

$$\boxed{W = \frac{1}{2} CV^2} \qquad\qquad\qquad \textbf{(12.5)}$$

For the network of Fig. 12.1, the capacitor will, for all practical purposes, charge up to E volts in *five* time constants, where a time constant (τ) is defined by

$$\boxed{\tau = RC} \qquad\qquad\qquad \textbf{(12.6)}$$

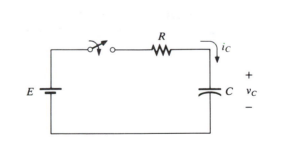

FIG. 12.1

In one time constant, the voltage v_C will charge up to 63.2% of its final value, in 2τ up to 86.5%, in 3τ up to 95.1%, in 4τ up to 98.1%, and in 5τ up to 99.3%, as defined by

$$v_C = E(1 - e^{-t/RC}) \qquad\qquad \textbf{(12.7)}$$

The current i_C is defined by

$$i_C = \frac{E}{R}(e^{-t/RC}) \qquad\qquad \textbf{(12.8)}$$

PROCEDURE

Part 1

(a) Construct the network of Fig. 12.2. Insert the measured resistor value. Be sure to note polarity on electrolytic capacitors as shown in the figure.

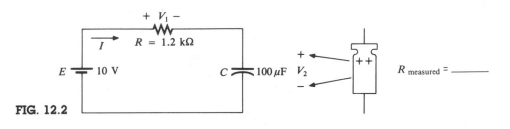

FIG. 12.2

(b) Calculate the steady-state value (defined by a period of time greater than five time constants) of the current I and the voltages V_1 and V_2.

$I =$ _____ , $V_1 =$ _____ , $V_2 =$ _____

(c) Measure the voltages V_1 and V_2 and calculate the current I from Ohm's law. Compare with the results of part 1(b) above.

$I =$ _____ , $V_1 =$ _____ , $V_2 =$ _____

(d) Calculate the energy stored by the capacitor.

$W =$ _____

(e) Carefully disconnect the supply and measure the voltage across the discon-
nected capacitor. Is there a reading? Why?

$V_C =$ _____

(f) Short the capacitor terminals with a lead. Explain any effect. Why was it
necessary to perform this step?

Part 2

(a) Construct the network of Fig. 12.3. Insert the measured resistor values.

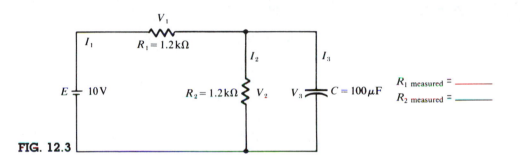

FIG. 12.3

(b) Using ideal elements, calculate the theoretical steady-state levels (time
greater than five time constants) of the following quantities. (The measured resistor values should be used.)

$I_1 =$ _____ , $I_2 =$ _____ , $I_3 =$ _____

$V_1 =$ _____ , $V_2 =$ _____ , $V_3 =$ _____

(c) Energize the system and measure the voltages V_1, V_2, and V_3. Calculate the
currents I_1 and I_2 from Ohm's Law and the current I_3 from Kirchhoff's current law. Compare the results with those of part 2(b).

$V_1 =$ _____ , $V_2 =$ _____ , $V_3 =$ _____

$I_1 =$ _____ , $I_2 =$ _____ , $I_3 =$ _____

Part 3

(a) Construct the network of Fig. 12.4. Insert the measured resistor values.

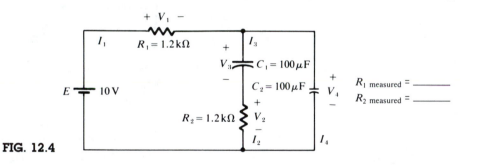

FIG. 12.4

(b) Assuming ideal elements, and using measured resistor values, calculate the theoretical steady-state levels of the following quantities:

$I_1 =$ _____ , $I_2 =$ _____ , $I_3 =$ _____ ,

$I_4 =$ _____

$V_1 =$ _____ , $V_2 =$ _____ , $V_3 =$ _____ ,

$V_4 =$ _____

(c) Energize the system and measure the voltages V_1, V_2, V_3, and V_4. Compare the results with those in part 3(b).

$V_1 = $ _____ , $V_2 = $ _____ , $V_3 = $ _____

$V_4 = $ _____

Part 4   Charging Network

(a) Calculate the time constant for the network of Fig. 12.5. Use the measured resistance value.

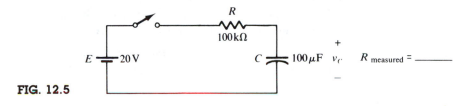

FIG. 12.5

$\tau = $ _____

(b) Discharge the capacitor and make sure the switch is in the open position. Then energize the source, close the switch, and note how many seconds pass before the voltage v_C reaches 63.2% of its final value or $(0.632)(20) = 12.64$ V (one time constant).

$t_{63.2\%} = $ _____

You will note that the t is somewhat larger than you expected from the calculations of part 4(a). This is due to the leakage resistance typically present in electrolytic capacitances of this capacitance level.

(c) An equivalent capacitance can be defined, however, that will permit an examination of the charging phase of a capacitor when applied to a dc source. It will be determined from

$$C_{eq} = \frac{t}{R} \qquad\qquad (12.9)$$

(where $t = $ time required to reach the 63.2% level and $R = $ measured value), as derived from the time constant equation $\tau = RC$.

Calculate the equivalent capacitance for the configuration of Fig. 12.5.

C_{eq_1} = _____

Repeat the above determination for the other capacitor.

C_{eq_2} = _____

From this point on, simply assume that each capacitor has a capacitance equivalent to C_{eq} and the data obtained for this experiment should be quite satisfactory.

(d) Determine the time constant and charging time for the network of Fig. 12.6. Use the measured resistance levels and C_{eq} for each capacitor.

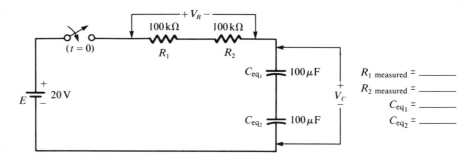

FIG. 12.6

τ = _____ , 5τ = _____

(e) Using a watch, record (to the best of your ability) the voltage across the resistors at the time intervals noted in Table 12.3 after the switch is closed. Complete the table for V_C using the fact that $V_C = E - V_R$.

TABLE 12.3

t (s)	5	10	15	20	25	30	35	40	45	50
V_R										
V_C										

(f) In one time constant, the voltage V_C should be 63.2% of its final steady-state value of 20 V. Is this verified by the data of part 4(e)?

(g) On Graph 12.1, plot the curves of v_R and v_C versus time.

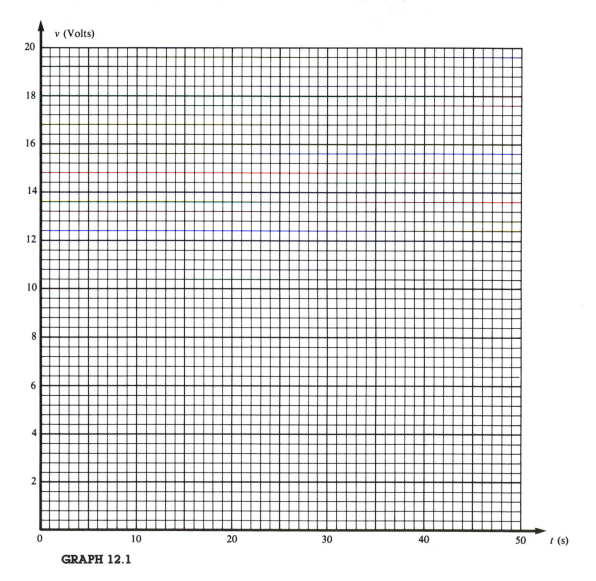

GRAPH 12.1

(h) Construct the network of Fig. 12.7. Insert the measured resistance level and each C_{eq}.

Repeat parts 4(d) through 4(g) for Fig. 12.7 and complete Table 12.4.

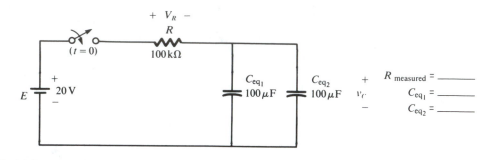

FIG. 12.7

$\tau =$ _____ , $5\tau =$ _____

TABLE 12.4

t (s)	10	20	30	40	50	60	70	80	90	100
V_R										
V_C										

On Graph 12.2, plot the curves of v_R and v_C versus time. Do the data verify the theoretical conclusions?

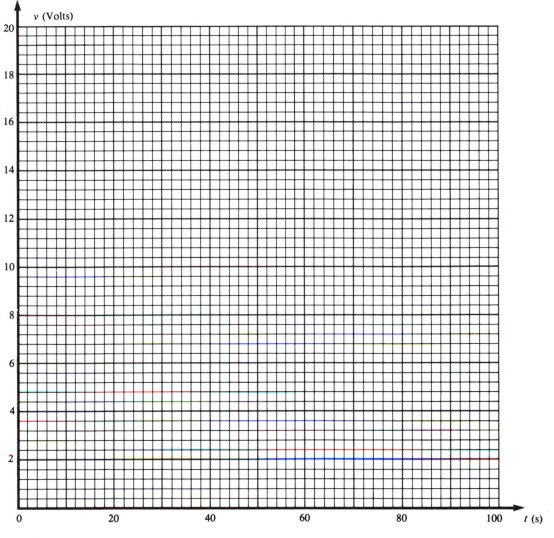

GRAPH 12.2

R-L and R-L-C Circuits with a dc Source Voltage

OBJECT

To investigate the response of R-L and R-L-C circuits to a dc voltage input.

EQUIPMENT REQUIRED

Resistors

1 — 10-Ω, 470-Ω, 1-kΩ, 2.2-kΩ

Inductors

1 — 10-mH

Capacitors

1 — 1-μF

Instruments

1 — DMM
1 — dc Power supply

Miscellaneous

1 — SPST switch
1 — NE-2 neon glow lamp

EQUIPMENT ISSUED

TABLE 13.1

Item	Manufacturer and Model No.	Laboratory Serial No.
DMM		
Power supply		

TABLE 13.2

Resistors	
Nominal Value	Measured Value
10 Ω	
470 Ω	
1 kΩ	
2.2 kΩ	

RÉSUMÉ OF THEORY

The inductor, like the capacitor, is an energy-storing device. The capacitor stores energy in the form of an electric field while the inductor stores it in the form of a magnetic field. The energy stored by an inductor (in joules) is given by

$$W = \frac{1}{2} LI^2 \tag{13.1}$$

In any circuit containing an inductor, the voltage across the inductor is determined by the inductance (L) and the rate of change of the current through the inductor:

$$v_L = L \frac{\text{change of current}}{\text{change of time}} = L \frac{\Delta I}{\Delta t} \tag{13.2}$$

If the current is constant (as in a dc circuit), $\Delta I/\Delta t = 0$, and the only voltage drop across the inductor is due to the dc resistance of the wire that makes up the inductor. Because an inductor has a dc resistance, we draw it schematically as shown in Fig. 13.1.

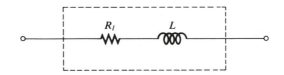

FIG. 13.1

The total inductance of inductors in series is given by

$$\boxed{L_T = L_1 + L_2 + L_3 + \cdots + L_N}$$ (13.3)

and in parallel by

$$\boxed{\frac{1}{L_T} = \frac{1}{L_1} + \frac{1}{L_2} + \frac{1}{L_3} + \cdots + \frac{1}{L_N}}$$ (13.4)

Note the correspondence of Eqs. (13.3) and (13.4) to the equations for resistive elements.

PROCEDURE

Part 1

Construct the circuit of Fig. 13.2. Insert the measured values of R_1 and R_l.

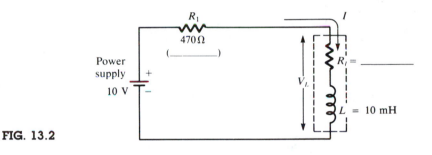

FIG. 13.2

(a) Calculate the current (I) and the voltage (V_L) for steady-state conditions. Since $5\tau = 5L/R = 5(10 \times 10^{-3})/470 \cong 0.1$ ms, steady-state conditions essentially exist once the network is constructed.

$I =$ _____ , $V_L =$ _____

(b) Measure V_L and determine I. (I is to be calculated by measuring the voltage across the 470-Ω resistor and using Ohm's Law.)

$I =$ _____ , $V_L =$ _____

Calculate R_l.

$R_l =$ _____

How does the value of R_l in part 1(b) compare with the measured value?

Is it possible for a practical inductor to have $R_l = 0$? Explain.

Part 2 Parallel R-L dc Circuit

Construct the circuit of Fig. 13.3. Insert the measured value of each resistor.

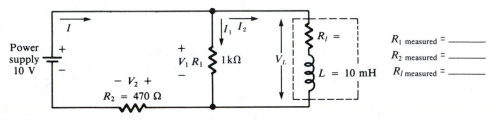

R_1 measured = _____

R_2 measured = _____

R_l measured = _____

FIG. 13.3

(a) Calculate I, I_1, and I_2 assuming an ideal inductor ($R_l = 0\ \Omega$). Use measured resistor values.

$I =$ _____ , $I_1 =$ _____ , $I_2 =$ _____

(b) Measure V_1 and V_2, and determine I, I_1, and I_2 using the equations

$$I = \frac{V_2}{R_2} \qquad I_1 = \frac{V_1}{R_1} \qquad I_2 = I - I_1$$

$V_1 =$ _____ , $V_2 =$ _____

$I =$ _____ , $I_1 =$ _____ , $I_2 =$ _____

How do the calculated and measured values for I, I_1, and I_2 compare?

Part 3   Series-Parallel *R-L-C* dc Circuit

Construct the network of Fig. 13.4. Insert the measured values of all resistors (including R_l).

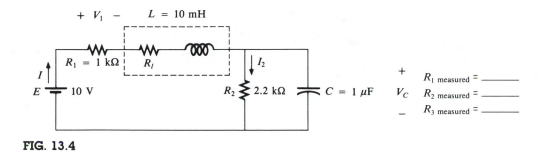

FIG. 13.4

R_1 measured = _____

R_2 measured = _____

R_3 measured = _____

(a) Using measured values, calculate I, I_2, and V_C.

$I =$ _____ , $I_2 =$ _____ , $V_C =$ _____

(b) Measure V_1 and V_C and calculate I and I_2 from the equations

$$I = V_1/R_1 \qquad I_2 = \frac{V_2}{R_2} = \frac{V_C}{R_2}$$

Compare the results with those in part 3(a).

$V_1 =$ _____ , $V_C =$ _____

$I =$ _____ , $I_2 =$ _____

(c) Calculate the energy stored by the capacitor and inductor.

$$W_C = \underline{\hspace{3cm}} \,, \quad W_L = \underline{\hspace{3cm}}$$

Part 4   Induced Voltage in an Inductor

(a) Place your power supply in series with the supply of another squad as shown in Fig. 13.5

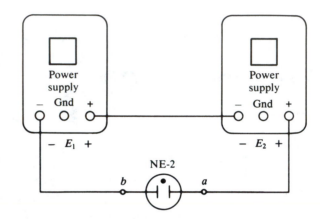

FIG. 13.5

Vary the voltages of both supplies until the lamp lights. Measure the voltage V_{ab} required to light the lamp, noting that $V_{ab} = E_1 + E_2$.

$$V_{ab} = \underline{\hspace{3cm}}$$

Repeat the above for the lamp of the other squad.

(b) Return the other supply and construct the circuit of Fig. 13.6.

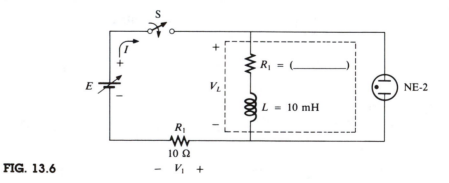

FIG. 13.6

Set the voltage source E to zero volts. Then close the switch and increase E until the voltage $V_L = 0.3$ V [significantly less than V_{ab} from part 4(a)]. Measure the voltage V_1 and calculate the current I delivered by the source to establish the magnetic field of the coil. Assume the bulb current is zero amperes.

$V_1 =$ _____ , $I =$ _____

Calculate the energy stored by the coil using $W_L = 1/2LI^2$.

$W_L =$ _____

Does the lamp light with the switch in the on (closed) position? Why not?

Now open the switch and note the effect on the lamp. Describe what happens.

Since $v_L = L(di/dt)$, the change in current di/dt was sufficiently high to develop a product $L(di/dt)$ that would equal the voltage V_{ab} measured earlier. Since the current dropped from the level I determined earlier to zero, $\Delta I = I - 0 = I$ in Eq. (13.2) and V_L must be at least V_{ab} volts. Substituting, we obtain

$$v_L = L \frac{\Delta I}{\Delta t}$$

$$V_{ab} = \frac{LI}{\Delta t}$$

and the switching time Δt can be determined from

$$\Delta t = \frac{LI}{V_{ab}}$$

Calculate Δt for your level of I and V_{ab}.

$\Delta t =$ _____

What is the effect on Δt of making the inductance larger? Comment accordingly.

An automobile ignition coil (an inductor) has an inductance of 5 H. The initial current is 2 A, and $\Delta t = 0.4$ ms. Calculate the induced voltage at the spark plug.

$V_L = $ _____

Design of a dc Ammeter and Voltmeter and Meter Loading Effects

OBJECT

To design and calibrate a dc ammeter and voltmeter.

EQUIPMENT REQUIRED

Resistors

1 — 10-Ω, 47-Ω

2 — 100-kΩ

1 — 0–10-kΩ potentiometer

Instruments

1 — VOM

1 — DMM

1 — dc Power supply

1 — 1-mA, 1000-Ω d'Arsonval meter movement

EQUIPMENT ISSUED

TABLE 14.1

Item	Manufacturer and Model No.	Laboratory Serial No.
VOM		
DMM		
Power supply		
Meter movement		

TABLE 14.2

Resistors	
Nominal Value	Measured Value
10 Ω	
47 Ω	
100 kΩ	
100 kΩ	

RÉSUMÉ OF THEORY

The basic meter movement is a current-sensitive device. Its construction consists of a coil suspended in a magnetic field with a pointer attached to it. When current is passed through the coil, an interaction of magnetic fields will cause the coil to rotate on its axis. The attached pointer will then indicate a particular angular displacement proportional to the magnitude of the current through the coil. D'Arsonval meter movements are delicate and sensitive dc milliammeters or microammeters used in the design and construction of dc voltmeters and ammeters.

A dc movement (d'Arsonval) can be used to construct a dc voltmeter by connecting a resistor in series with the movement, as shown in Fig. 14.1(a). This series resistor is called the *multiplier*. An ammeter (Fig. 14.1b) is constructed by placing a resistor in parallel with the meter movement. This resistor is called the *shunt* resistor.

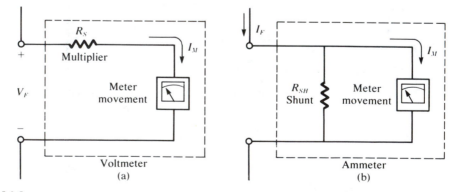

FIG. 14.1

The value of R_S can be calculated as follows:

$$R_S = \frac{V_F}{I_M} - R_M$$

The value of R_{SH} can be calculated as follows:

$$R_{SH} = \left(\frac{I_M}{I_F - I_M}\right)(R_M) \quad \text{(14.1)}$$

where: R_S = multiplier resistor

R_{SH} = shunt resistor

I_M = full-scale deflection current of the meter movement

R_M = internal resistance of the meter movement

V_F = maximum value of the voltage to be measured for the range being designed

I_F = maximum value of the current to be measured for the range being designed

EXAMPLE   Using a 0–5-mA meter movement with an internal resistance of 5 kΩ, design

    (a) a 0–50-V voltmeter

    (b) a 0–100-mA milliammeter

    Given I_M = 5 mA and R_M = 5 kΩ, then

Solutions

    (a) $R_S = \dfrac{V_F}{I_M} - R_M = \dfrac{50}{5 \times 10^{-3}} - 5 \times 10^3 = 5 \text{ k}\Omega$

(note Fig. 14.2a) and

    (b)

$$R_{SH} = \left(\frac{I_M}{I_F - I_M}\right)(R_M) = \left[\frac{5 \times 10^{-3}}{(100 \times 10^{-3}) - (5 \times 10^{-3})}\right](5 \times 10^3) = 263 \text{ }\Omega$$

(note Fig. 14.2b).

    The current sensitivity of a meter is a measure of the current necessary to obtain a full-scale deflection of the meter movement. The inverse of current sensitivity of a voltmeter is the ohm/volt sensitivity of the meter. Therefore, the dc multirange voltmeter, with a 1000-ohm/volt sensitivity, has a current sensitivity of 1/1000 or 1 mA. The VOM, with a 20,000-ohm/volt sensitivity for dc, has a current sensitivity of 1/20,000 or 50 μA.

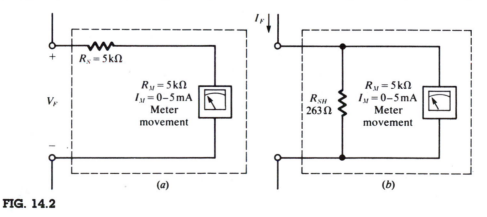

(a)   (b)

FIG. 14.2

Ideally, the internal resistance of a voltmeter should be infinite to minimize its impact on a circuit when taking measurements. As shown in Fig. 14.3, the internal resistance of the meter will appear in parallel with the element under investigation. The effect is to reduce the net resistance of the network since $R_2 > R_2 \| R_{int}$, causing V_2 to decrease in magnitude. The closer R_{int} is to R_2, the more it will affect the reading. As a rule, R_{int} should be more than 10 times R_2 to ensure a reasonable level of accuracy in the reading. The internal resistance of a voltmeter is the series combination of the multiplier and movement resistances, as shown in Fig. 14.3.

The accuracy in percent of the full-scale reading indicates the maximum error that will occur in the meter. The Simpson 377 dc voltmeter has an accuracy of $\pm 5\%$ of full scale. Thus, when set on the 100-V scale, the reading may be in error by ± 5 V at any point on the scale. For maximum accuracy, therefore, always choose a scale setting that will give a reading (meter deflection) as close to full scale as possible.

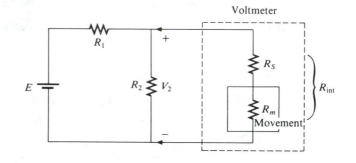

FIG. 14.3

PROCEDURE

Part 1   Design of a dc Voltmeter

(a) Use the 1000-Ω, 1-mA meter movement to design a 0–10-V dc voltmeter using the appropriate variable resistor. Ask your instructor to check your design before proceeding. Show all calculations.

$R_S =$ _____

Calibration of the Designed Voltmeter

(b) Connect your newly constructed voltmeter in parallel with a voltage source and the DMM as shown in Fig. 14.4. Set the variable resistor to R_S with your DMM in the ohmmeter mode.

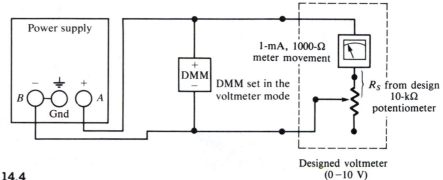

FIG. 14.4

Designed voltmeter
(0–10 V)

Now vary the power supply from zero to 10 V in 1-V steps and record your readings in Table 14.3.

TABLE 14.3

DMM	Designed Voltmeter
1 V	
2 V	
3 V	
4 V	
5 V	
6 V	
7 V	
8 V	
9 V	
10 V	

Calculate the percent difference of the reading of your voltmeter at three different readings: 1, 5, and 10 V. Use

$$\% \text{ Difference} = \left[\frac{\left| \left(\begin{array}{c} \text{designed voltmeter} \\ \text{reading} \end{array} \right) - \left(\begin{array}{c} \text{DMM} \\ \text{reading} \end{array} \right) \right|}{\text{DMM reading}} \right] \times 100\%$$

% difference (at 1 V) = _____

% difference (at 5 V) = _____

% difference (at 10 V) = _____

What do you conclude from the above results?

What is the sensitivity of your voltmeter?

Part 2   Design of a dc Milliammeter

Use the meter movement (1-mA, 1000-Ω) to design a 100-mA milliammeter. Let your instructor check your design before proceeding. Then connect the milliammeter into the circuit shown in Fig. 14.5 and compare its reading with that of the multirange dc milliammeter set on the proper scale. Start with the power supply set to zero and increase it so that the current I varies from zero to 100 mA in 10-mA steps using the DMM in the milliammeter mode. Record the data in Table 14.4. Show your calculations.

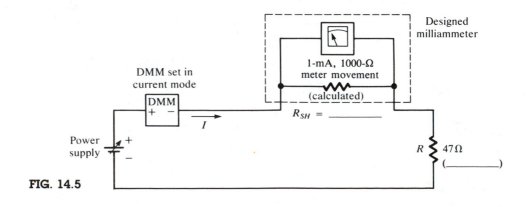

FIG. 14.5

$R_{SH} = $ _____

TABLE 14.4

DMM	Designed Milliammeter
10 mA	
20 mA	
30 mA	
40 mA	
50 mA	
60 mA	
70 mA	
80 mA	
90 mA	
100 mA	

Calculate the percent difference of the reading of your milliammeter at three different readings: 10, 50, and 100 mA. Use the equation provided in part 1.

% difference (at 10 mA) = _____

% difference (at 50 mA) = _____

% difference (at 100 mA) = _____

What conclusions do you draw from the above results?

Part 3   Meter Loading Effects

(a) Construct the circuit of Fig. 14.6.

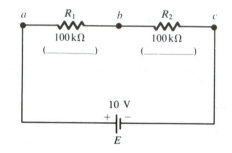

FIG. 14.6

What is the theoretical value of the voltage V_{ab}? Use measured resistor values.

V_{ab} = _____

(b) Read the voltage V_{ab} using the DMM and the VOM on the 10-V scale.

V_{ab}(DMM) = _____ , V_{ab}(VOM) = _____

(c) Most DMMs have an internal impedance of 10 MΩ or greater and can be considered an open circuit across the 100-kΩ resistors. The VOM, however, has an ohm/volt rating of 20,000 and therefore an internal resistance of (10 V)(20,000 Ω/V) = 200 kΩ on the 10-V scale. Using this internal resistance value for the VOM, calculate the voltage V_{ab} and compare with the result of part 3(b) for the VOM.

V_{ab}(calculated − VOM) = _____ ,

V_{ab}[VOM from part 3(b)] = _____

(d) Will the 50-V scale of the VOM give a better reading? Record below and note any disadvantages of using this higher scale.

V_{ab}(50-V scale) = _____

(e) If the resistors were replaced by 1-kΩ values, would there be an improvement in the readings of the VOM? Show why with a sample calculation.

Wheatstone Bridge and Δ-Y Conversions

OBJECT

To become familiar with the Wheatstone bridge and Δ-Y conversions.

EQUIPMENT REQUIRED

Resistors

1 — 91-Ω

2 — 200-Ω

3 — 330-Ω, 1-kΩ

1 — 0–1-kΩ potentiometer

1 — Unmarked fixed resistor in the range 47 Ω to 220 Ω

Instruments

1 — DMM (or VOM)

1 — dc Power supply

1 — Commercial Wheatstone bridge (if available)

EQUIPMENT ISSUED

TABLE 15.1

Item	Manufacturer and Model No.	Laboratory Serial No.
DMM (or VOM)		
Power supply		
Wheatstone bridge		

Since more than one resistor of a particular value will appear in some configurations of this experiment, the measured value of each should be determined just prior to inserting them into the network.

RÉSUMÉ OF THEORY

The Wheatstone bridge is an instrument used to make precision measurements of unknown resistance levels. The basic configuration appears in Fig. 15.1. The unknown resistance is R_x, and R_1, R_2, and R_3 are precision resistors of known value.

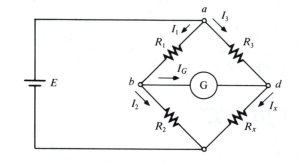

FIG. 15.1

The network is balanced when the galvanometer (G) reads zero.

We know from circuit theory that if $I_G = 0$ A, the voltage V_{bd} is zero, and

$$V_{ab} = V_{ad} \quad \text{and} \quad V_{bc} = V_{dc}$$

By substitution,

$$I_1 R_1 = I_3 R_3 \tag{15.1}$$

and

$$I_2 R_2 = I_x R_x \tag{15.2}$$

Solving Eq. (15.1) for I_1 yields

$$I_1 = \frac{I_3 R_3}{R_1}$$

Substituting I_1 for I_2 and I_3 for I_x in Eq. (15.2) (since $I_1 = I_2$ and $I_3 = I_x$ with $I_G = 0$ A), we have

$$I_1 R_2 = I_3 R_x \quad \text{or} \quad \left(\frac{I_3 R_3}{R_1}\right) R_2 = I_3 R_x$$

Canceling I_3 from both sides and solving for R_x, we obtain

$$R_x = \frac{R_2}{R_1} R_3 \qquad (15.3)$$

or, in the ratio form,

$$\boxed{\frac{R_1}{R_2} = \frac{R_3}{R_x}} \qquad (15.4)$$

In the commercial Wheatstone bridge, R_1 and R_2 are variable in decade steps so that the ratio R_1/R_2 is a decimal or integral multiplier. R_3 is a continuous variable resistor, such as a slide-wire rheostat. Before the unknown resistor is connected to the terminals of the commercial bridge, the R_1/R_2 ratio (called the *factor of the ratio arms*) is adjusted for that particular unknown resistor. After the resistor is connected, R_3 is adjusted until there is no detectable current indicated by the galvanometer. (Galvanometer sensitivities are usually 10^{-10} A or better.) The unknown resistance value is the ratio factor times the R_3 setting.

There are certain circuit configurations in which the resistors do not appear to be in series or parallel. Under these conditions, it is necessary to convert the circuit in question from one form to another. The two circuits to be investigated in this experiment are the delta (Δ) and the wye (Y), both of which appear in Fig. 15.2. To convert a Δ to a Y (or vice versa), we use the following conversion equations:

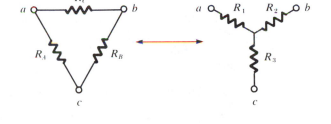

FIG. 15.2

$$R_1 = \frac{R_A R_C}{R_A + R_B + R_C} \qquad R_2 = \frac{R_B R_C}{R_A + R_B + R_C} \qquad R_3 = \frac{R_A R_B}{R_A + R_B + R_C} \qquad (15.5)$$

$$R_A = \frac{R_1 R_2 + R_1 R_3 + R_2 R_3}{R_2} \qquad R_B = \frac{R_1 R_2 + R_1 R_3 + R_2 R_3}{R_1}$$

$$R_C = \frac{R_1 R_2 + R_1 R_3 + R_2 R_3}{R_3} \qquad (15.6)$$

If $R_A = R_B = R_C$,

$$\boxed{R_Y = \frac{R_\Delta}{3}} \qquad (15.7)$$

If $R_1 = R_2 = R_3$,

$$\boxed{R_\Delta = \frac{R_Y}{3}}$$ (15.8)

PROCEDURE

Part 1   Wheatstone Bridge Network

(a) Construct the network of Fig. 15.3. Insert the measured value of each resistor and set the potentiometer to the maximum resistance setting.

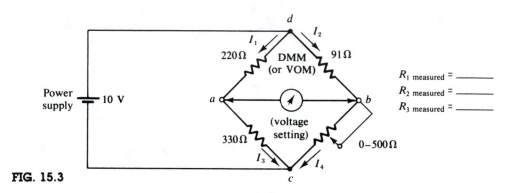

FIG. 15.3

(b) Starting with the meter on a higher voltage scale, vary the potentiometer until the voltage V_{ab} is as close to zero as possible. Then drop the voltage scales to the lowest range possible to set the voltage V_{ab} as close to zero volts as possible. The bridge is now balanced.

(c) Measure the voltages V_{da}, V_{db}, V_{ac}, and V_{bc}.

$V_{da} = $ _____ , $V_{db} = $ _____ , $V_{ac} = $ _____ ,

$V_{bc} = $ _____

(d) Calculate the currents I_1 and I_3 using Ohm's Law. Are they equal as defined in the Résumé of Theory?

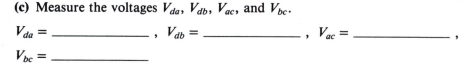

$I_1 = $ _____ , $I_3 = $ _____

(e) Disconnect one lead of the potentiometer (used as a rheostat) and measure its resistance.

$R_{\text{pot}} = R_x = $ _____

Calculate the currents I_2 and I_x using the results of part 1(c) and Ohm's Law. Are they equal as defined in the Résumé of Theory?

$I_2 = $ _____ , $I_x = $ _____

(f) Verify that the following ratio is satisfied:

$$\frac{R_1}{R_3} = \frac{R_2}{R_x}$$

(g) Replace the 91-Ω resistor by the unknown resistor. Proceed as before to adjust the potentiometer until $V_{ab} \cong 0$ V. Remove the variable resistor and measure its resistance with the ohmmeter section of your multimeter.

$R_{pot} = $ _____

(h) Calculate the unknown resistance using Eq. (15.4).

$R_x = $ _____

(i) What is the maximum value of resistance that this network could measure? Why?

$R_{max} = $ _____

Part 2   Commercial Wheatstone Bridge

(a) Use the commercial Wheatstone bridge to measure the resistance of the unknown resistor.

$R_x = $ _____

(b) Measure R_x using the ohmmeter section of your multimeter. Compare with the results of part 2(a) and part 1(h).

$R_x = $ _____

Part 3   Δ-Y Conversions

(a) Construct the network of Fig. 15.4. Insert the measured values of the 220-Ω resistors. Assume for the moment that each 1-kΩ resistor is exactly 1 kΩ.

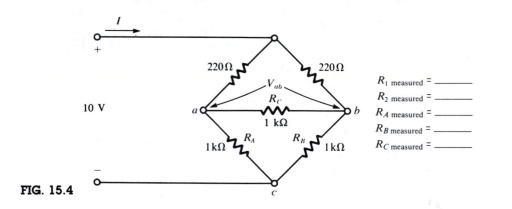

FIG. 15.4

(b) Calculate the current I and the voltage V_{ab} using any method other than a Δ-Y conversion (that is, mesh analysis, nodal analysis, and so on).

$I =$ _____ , $V_{ab} =$ _____

(c) Measure the current I and the voltage V_{ab} and compare to the results of part 3(b).

$I =$ _____ , $V_{ab} =$ _____

(d) Calculate the equivalent Y for the Δ formed by the three 1-kΩ resistors. Draw the equivalent circuit with the Δ replaced by the Y. Insert the values of the resistors in the Y in Fig. 15.5 and also indicate on the diagram those available fixed resistors that have a resistance level closest to the calculated value.

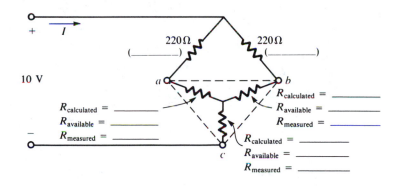

FIG. 15.5

(e) Construct the network drawn in part 3(d) and measure the current I and the voltage V_{ab}. Are they approximately the same as those obtained in part 3(c)? If so, why?

$I =$ _____ , $V_{ab} =$ _____

(f) Calculate the input resistance to the network of part 3(d) using the measured resistor values.

R_T(calculated) = _____

(g) Disconnect the supply and measure the input resistance to the network of (d). Compare with the results of part 3(f).

R_T(measured) = _____

(h) Determine the input resistance to the network of part 3(a) using $R_T = E/I$ and compare with the results of part 3(g). Should they compare? Why?

R_T = _____

(i) Calculate I using the result of part 3(f) and compare with the value measured in part 3(e).

I = _____

Math Review

ac

The analysis of sinusoidal ac networks will require the extended use of the Pythagorean theorem, trigonometric functions, and vector operations. This review will begin to establish the mathematical foundation necessary to work with each of these functions and operations in a confident, correct, and accurate manner.

It is assumed that a calculator or table is available to determine (with reasonable accuracy) the value of the trigonometric functions (sine, cosine, tangent) and the square and square root of a number.

PYTHAGOREAN THEOREM

The Pythagorean theorem states that the square of the hypotenuse of a right triangle is equal to the sum of the squares of the other sides of the triangle. With the notation in Fig. M.1, the theorem has the following form:

$$Z^2 = X^2 + Y^2 \qquad \text{(M.1)}$$

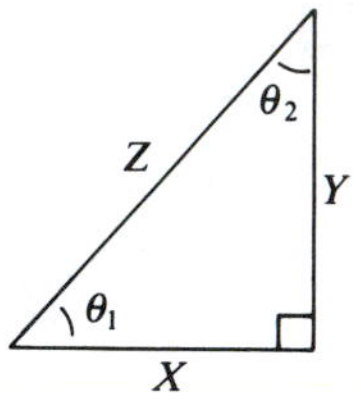

FIG. M.1

For θ_1,

$$\sin \theta_1 = \frac{\text{opposite}}{\text{hypotenuse}} = \frac{Y}{Z} \qquad \text{(M.2)}$$

$$\cos \theta_1 = \frac{\text{adjacent}}{\text{hypotenuse}} = \frac{X}{Z} \qquad \text{(M.3)}$$

$$\tan \theta_1 = \frac{\sin \theta_1}{\cos \theta_1} = \frac{Y/Z}{X/Z} = \frac{Y}{X} = \frac{\text{opposite}}{\text{adjacent}} \qquad \text{(M.4)}$$

For θ_2,

$$\sin \theta_2 = \frac{X}{Z}$$

$$\cos \theta_2 = \frac{Y}{Z}$$

$$\tan \theta_2 = \frac{X}{Y}$$

To determine θ_1 and θ_2, we use

$$\theta_1 = \sin^{-1} \frac{Y}{Z} = \cos^{-1} \frac{X}{Z} = \tan^{-1} \frac{Y}{X}$$

$$\theta_2 = \sin^{-1} \frac{X}{Z} = \cos^{-1} \frac{Y}{Z} = \tan^{-1} \frac{X}{Y} \qquad \text{(M.5)}$$

EXAMPLE 1   Find the hypotenuse Z and the angle θ_1 for the right triangle of Fig. M.2.

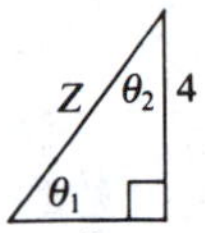

FIG. M.2

Solution

$$Z = \sqrt{X^2 + Y^2} = \sqrt{(3)^2 + (4)^2} = \sqrt{25} = 5$$

and

$$\theta_1 = \tan^{-1}\frac{4}{3} = \tan^{-1} 1.333 \ldots \cong 53.13°$$

or

$$\theta_1 = \sin^{-1}\frac{4}{5} = \sin^{-1} 0.8 \cong 53.13°$$

or

$$\theta_1 = \cos^{-1}\frac{3}{5} = \cos^{-1} 0.6 \cong 53.13°$$

EXAMPLE 2   Determine Y and θ_1 for the right triangle of Fig. M.3.

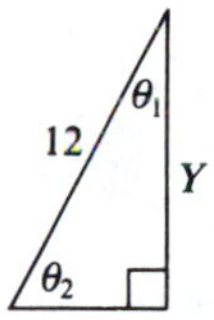

FIG. M.3

Solution

$$Z^2 = X^2 + Y^2$$

or

$$Y^2 = Z^2 - X^2$$

and

$$Y = \sqrt{Z^2 - X^2} \qquad\qquad \textbf{(M.6)}$$

Substituting,

$$Y = \sqrt{(12)^2 - (6)^2} = \sqrt{108} \cong 10.39$$

and

$$\theta_1 = \sin^{-1}\frac{X}{Z} = \sin^{-1}\frac{6}{12} = \sin^{-1} 0.5 = 30°$$

PROBLEMS

1. Determine θ_2 for Fig. M.2.

    $\theta_2 =$ _____

2. Determine θ_2 for Fig. M.3.

    $\theta_2 =$ _____

3. **a.** Determine X for the right triangle of Fig. M.4.

    $X =$ _____

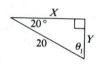

FIG. M.4

    b. Determine θ_1 and θ_2 for Fig. M.4.

    $\theta_1 =$ _____ , $\theta_2 =$ _____

4. Refer to Fig. M.5.

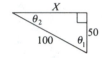

FIG. M.5

    a. Determine Y.

    $Y =$ _____

    b. Find X.

    $X =$ _____

    c. Determine θ_1.

    $\theta_1 =$ _____

VECTOR REPRESENTATION

There are two common forms for representing vectors with the tail fixed at the origin. The *rectangular* form has the appearance shown in Fig. M.6, where the real and imaginary components define the location of the head of the vector.

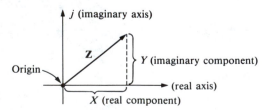

FIG. M.6

The imaginary component is defined by the component of the vector that has the *j* associated with it, as appearing in the following representation of the rectangular form:

Rectangular Form:

$$\boxed{\mathbf{Z} = X + jY}$$
(M.7)

In the first and fourth quadrant, X is positive, while in the second and third quadrant, it is negative. Similarly, the imaginary component is positive in the first and second quadrants and negative in the third and fourth.

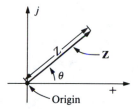

FIG. M.7

The *polar* form has the appearance shown in Fig. M.7, where the angle θ is *always* measured from the positive real axis. The equation form is as follows:

Polar Form:

$$\boxed{\mathbf{Z} = Z\,\angle\theta}$$
(M.8)

Through Fig. M.8, it should be fairly obvious that the two forms are related by the following equations.

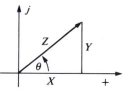

FIG. M.8

Rectangular to Polar:

$$Z = \sqrt{X^2 + Y^2}$$
(M.9)

$$\theta = \tan^{-1}\frac{Y}{X}$$
(M.10)

Polar to Rectangular:

$$X = Z\cos\theta$$
(M.11)

$$Y = Z\sin\theta$$
(M.12)

EXAMPLE 3   Determine the rectangular and polar forms of the vector of Fig. M.9.

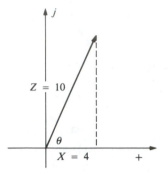

FIG. M.9

Solution   Rectangular form:

$$Y = \sqrt{Z^2 - X^2} = \sqrt{(10)^2 - (4)^2} = \sqrt{84} = 9.165$$

and

$$\mathbf{Z} = 4 + j9.165$$

Polar form:

$$\theta = \tan^{-1}\frac{Y}{X} = \tan^{-1}\frac{9.165}{4} = \tan^{-1} 2.291 = 66.42°$$

and

$$\mathbf{Z} = 10\ \angle 66.42°$$

EXAMPLE 4   Determine the polar form of the vector of Fig. M.10.

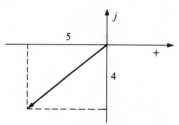

FIG. M.10

Solution   In conversions of this type, it is usually best to isolate a right triangle, as shown in Fig. M.11, and determine θ' and Z.

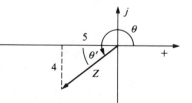

FIG. M.11

The angle θ' is not the angle θ of the polar form but it can be determined from the right triangle and then used to calculate θ.

In this case,

$$Z = \sqrt{X^2 + Y^2} = \sqrt{(5)^2 + (4)^2} = \sqrt{41} \cong 6.4$$

and

$$\theta' = \tan^{-1} \frac{4}{5} = \tan^{-1} 0.8 = 38.66°$$

Using Fig. M.11, we find

$$\theta = 180° + \theta' = 180° + 38.66° = 218.66°$$

and

$$\mathbf{Z} = 6.4 \angle 218.66°$$

PROBLEMS

5. Determine the polar form of each of the following vectors:

   a. $\mathbf{Z} = 10 + j10 = $ _____

   b. $\mathbf{Z} = -2 + j8 = $ _____

   c. $\mathbf{Z} = -50 - j200 = $ _____

   d. $\mathbf{Z} = +0.4 - j0.8 = $ _____

6. Determine the rectangular form of each of the following vectors:

   a. $\mathbf{Z} = 6 \angle 37.5° = $ _____

   b. $\mathbf{Z} = 2 \times 10^{-3} \angle 100° = $ _____

   c. $\mathbf{Z} = 52 \angle -120° = $ _____

   d. $\mathbf{Z} = 1.8 \angle -30° = $ _____

ADDITION AND SUBTRACTION OF VECTORS

The sum or difference of vectors is normally found in the rectangular form. The operation can be performed in the polar form only if the vectors have the same angle or are out of phase by 180°.

Addition

Rectangular Form:

$$(X_1 + jY_1) + (X_2 + jY_2) = (X_1 + X_2) + j(Y_1 + Y_2) \tag{M.13}$$

Polar Form:

$$Z_1 \angle\theta + Z_2 \angle\theta = (Z_1 + Z_2) \angle\theta \tag{M.14}$$

Subtraction

Rectangular Form:

$$(X_1 + jY_1) - (X_2 + jY_2) = (X_1 - X_2) + j(Y_1 - Y_2) \qquad \text{(M.15)}$$

Polar Form:

$$Z_1 \angle\theta - Z_2 \angle\theta = (Z_1 - Z_2) \angle\theta \qquad \text{(M.16)}$$

EXAMPLE 5   Determine the sum of the following vectors:

$$X_1 = 4 + j4 \qquad X_2 = 6 \angle 120°$$

Solution   Using Fig. M.12 to convert X_2 to rectangular form,

$$Y = 6 \sin 60° = 6(0.866) = 5.196$$
$$X = 6 \cos 60° = 6(0.5) = 3$$

and

$$Z = 6 \angle 120° = -3 - j5.196$$

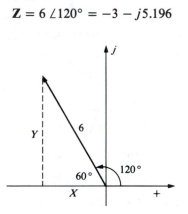

FIG. M.12

Note in the above conversion that the proper sign for X and Y is simply determined by the location of the vector.

$$X_{sum} = X_1 + X_2 = (4 + j4) + (-3 + j5.196)$$

Note the use of parentheses around each vector to identify clearly each vector and the sign of each term.

$$X_{sum} = (4 - 3) + j(4 + 5.196) = 1 + j9.196$$

EXAMPLE 6   Determine the difference between the following vectors:

$$X_1 = -5 - j4 \qquad X_2 = 3 - j8$$

Solution

$$X_{difference} = X_1 - X_2 = (-5 - j4) - (3 - j8) = (-5 - 3) + j(-4 + 8)$$

Carefully review the sign of each term of the above equation and note how the parentheses were an aid in generating the proper signs for the following solution:

$$X_{difference} = -8 + j4$$

PROBLEMS

7. Perform the following vector operations:

    a. $(10 + j30) + (-4 - j8) = $ _____

    b. $(10 \angle 45°) - (7.07 - j7.07) = $ _____

    c. $(50 \angle 60°) + (0.8 \angle 60°) = $ _____

    d. $(16 \angle -20°) + (14 \angle 160°) = $ _____

    e. $(5000 + j1000) - (2500 \angle -60°) = $ _____

    f. $(5 \angle 0°) + (20 \angle -90°) - (6 \angle 180°) = $ _____

MULTIPLICATION AND DIVISION OF VECTORS

Multiplication

Multiplication of vectors is usually performed in the polar form.

Polar Form:

$$\boxed{(Z_1 \angle \theta_1)(Z_2 \angle \theta_2) = Z_1 Z_2 \; \underline{/\theta_1 + \theta_2}}$$ **(M.17)**

Division

The division of vectors is normally performed in the polar form.

Polar Form:

$$\boxed{\frac{Z_1 \angle \theta_1}{Z_2 \angle \theta_2} = \frac{Z_1}{Z_2} \; \underline{/\theta_1 - \theta_2}}$$ **(M.18)**

EXAMPLE 7   Perform the following operation.

$$\frac{\mathbf{X}_1}{\mathbf{X}_2} \text{ if } \mathbf{X}_1 = 80 \angle 10° \quad \text{and} \quad \mathbf{X}_2 = 40 \angle -20°$$

Solution

$$\mathbf{X}_{\text{division}} = \frac{\mathbf{X}_1}{\mathbf{X}_2} = \frac{80 \angle 10°}{40 \angle -20°} = \frac{80}{40} \; \underline{/10 - (-20°)} = 2 \; \underline{/10 + 20°}$$

$$= 2 \angle 30°$$

PROBLEMS

8. Perform the following operations:

   a. $(50 \angle 60°)(4 \angle 80°) = \underline{\hspace{2cm}}$

   b. $(0.8 \angle -10°)(3 \angle 30°)(-4 \angle 20°) = \underline{\hspace{2cm}}$

   c. $(5 + j8)(4 - j2) = \underline{\hspace{2cm}}$

   d. $(48 \angle 20°)(10 - j1) = \underline{\hspace{2cm}}$

   e. $(0.9 - j0.5)(40 + j10) = \underline{\hspace{2cm}}$

   f. $(4000 \angle 0°)(2000 \angle -20°)(1000 + j1000) = \underline{\hspace{2cm}}$

9. Perform the following operations:

   a. $\dfrac{1000 \angle 60°}{20 \angle -50°} = \underline{\hspace{2cm}}$

   b. $\dfrac{0.008 \angle 50°}{5 \times 10^{-6} \angle 100°} = \underline{\hspace{2cm}}$

   c. $\dfrac{6 + j6}{2 + j2} = \underline{\hspace{2cm}}$

   d. $\dfrac{10 \angle 30°}{4 + j10} = \underline{\hspace{2cm}}$

   e. $\dfrac{20{,}000 + j10{,}500}{5000 \angle -50°} = \underline{\hspace{2cm}}$

   f. $\dfrac{(50 \angle 20°)(20 \angle -30°)}{(5 + j5)(4 \angle -10°)} = \underline{\hspace{2cm}}$

The Oscilloscope

OBJECT

To introduce the basic components and use of an oscilloscope and audio oscillator (or function generator).

EQUIPMENT REQUIRED

Instruments

1 — DMM (or VOM)

1 — Oscilloscope

1 — Audio oscillator or function generator

Miscellaneous

2 — D batteries with holders

EQUIPMENT ISSUED

TABLE 1.1

Item	Manufacturer and Model No.	Laboratory Serial No.
DMM (or VOM)		
Oscilloscope		
Audio oscillator or function generator		

RÉSUMÉ OF THEORY

Oscilloscope

The cathode-ray oscilloscope is an instrument that displays the variations of a voltage with time on a *cathode-ray tube* (CRT) such as that appearing in Fig. 1.1. The CRT is an evacuated glass envelope, shaped as shown in Fig. 1.1, with the following essential components:

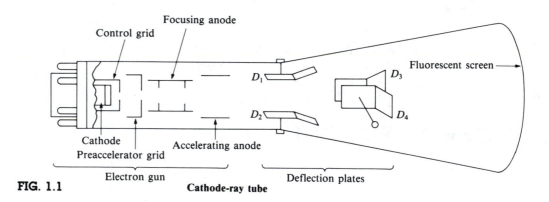

FIG. 1.1   **Cathode-ray tube**

1. An electron gun that generates rapidly moving electrons, shapes them into a pencil-like beam, and then directs them along the axis of the tube
2. Deflection plates for producing a vertical or horizontal deflection of the beam from its original path
3. A fluorescent screen that emits light at the point where the beam strikes

The *electron gun* is made up of the following components:

1. A heater and cathode that serve as the source of the electrons
2. A control electrode that serves to vary the strength of the beam current
3. A focusing electrode that focuses the beam to a sharp point on the screen
4. Accelerating and preaccelerating electrodes that give the electrons the high velocity required to reach the screen and cause secondary emission

The deflection of the beam may be accomplished either electrostatically or electromagnetically. Most oscilloscopes use electrostatic deflection (Fig. 1.1). To deflect the beam, a potential is applied across plates D_1 and D_2, or D_3 and D_4. The horizontal plates D_1 and D_2 deflect the beam vertically, whereas the vertical plates D_3 and D_4 deflect the beam horizontally. The deflection of the beam is directly proportional to the impressed voltage.

The CRT is, in itself, not a very sensitive device. A 5-in. CRT, under ordinary conditions, will give about 1 in. of deflection for a difference of potential of 100 V. The signals encountered most frequently are usually well under 100 V, so it becomes necessary to amplify the signals before a usable deflection will be obtained. Two deflection amplifiers are required, one for each set of plates.

For the oscilloscope to indicate the variations of an electrical quantity in the vertical direction as a function of time, there must be a voltage impressed on the horizontal deflection plates that varies linearly with time. This voltage is shown graphically in Fig. 1.2. An oscillator that can generate such a voltage is called, for obvious reasons, a *sawtooth* oscillator. Frequently, the term *sweep* oscillator is used. Figure 1.3 provides a block diagram of a general-purpose oscilloscope.

FIG. 1.2

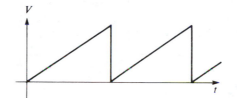

FIG. 1.3

All of the controls for the proper operation of an oscilloscope are mounted on the front panel of the instrument. Figure 1.4 indicates the approximate locations of the controls found on most general-purpose oscilloscopes. The locations of the controls shown vary according to manufacturer. Table 1.2 describes the function of each.

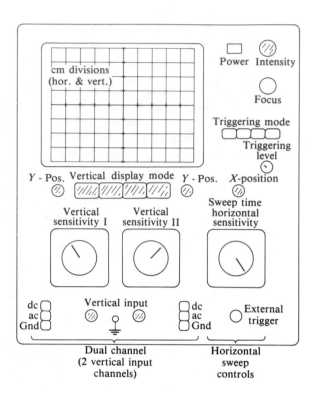

FIG. 1.4

TABLE 1.2

Control	Function
Power	Turns on the main power.
Intensity	Controls the intensity of the pattern on the screen.
Focus	Focuses the electron beam so that the pattern will be clearly defined.
Triggering mode	Determines the type of triggering for the horizontal sweeping pattern.
Triggering level	Determines the level at which triggering should occur.
Vertical display mode	Determines whether one or two signals will be displayed at the same time and which technique will be used to display the signals.
Y-position	Controls the vertical location of the pattern.
X-position	Controls the horizontal location of the pattern.
Vertical sensitivity	Determines the volts/cm for the vertical axis of display.
Sweep time horizontal sensitivity	Determines the time/cm for the horizontal axis of display.
dc/ac/Gnd switch	Determines whether dc levels will be displayed on the screen and permits determining the Gnd (zero-volt input level) of the display.

(a) Voltage Measurements

dc Levels:

To use the scope to measure dc levels, first place the dc/ac/Gnd switch in the Gnd position to establish the ground (zero-volt) level on the screen.

Then switch the dc/ac/Gnd switch to the dc position to measure the dc level. In the ac mode, a capacitor blocks the dc from the screen.

Then place the scope leads across the unknown dc level and use the following equation to determine the dc level:

$$\text{dc Level (V)} = \text{deflection (cm)} \times \text{vertical sensitivity (V/cm)} \qquad \textbf{(1.1)}$$

ac Levels:

After re-establishing the ground level, place the dc/ac/Gnd switch in the ac mode and connect the scope leads across the unknown voltage. The peak-to-peak voltage can then be determined from

$$V_{p\text{-}p}(\text{V}) = \text{deflection peak to peak (cm)} \times \text{vertical sensitivity (V/cm)} \qquad \textbf{(1.2)}$$

(b) Frequency Measurements The oscilloscope can be used to set the frequency of an audio oscillator or function generator using the *horizontal sensitivity* in the following manner.

Determine the period of the desired waveform and then calculate the number of divisions required to display the waveform on the horizontal axis using the provided μs/cm, ms/cm, or s/cm on the horizontal sensitivity control. Then adjust the audio oscillator or function generator to provide the proper horizontal deflection for the desired frequency.

Of course, the reverse of the above procedure will determine the frequency of an unknown signal.

Audio Oscillator and Function Generator

The *audio oscillator* is designed to provide a sinusoidal waveform in the frequency range *audible* by the human ear. A *function generator* typically expands on the capabilities of the audio oscillator by providing a square wave and triangular waveform with an increased frequency range. Either instrument is suitable for this experiment since we will be dealing only with sinusoidal waveforms in the audio range.

Most oscillators and generators require that the magnitude of the output signal be set by an oscilloscope or DMM (or VOM). That is, the amplitude dial of the oscillator is not graduated and the peak or peak-to-peak value is set by connecting the output of the oscillator to a scope or meter and adjusting the amplitude dial until the desired voltage is obtained.

PROCEDURE

Part 1 Introduction

(a) Your instructor will introduce the basic operation of the oscilloscope and audio oscillator or function generator.

(b) Turn on the oscilloscope and establish a horizontal line centered on the face of the screen. There are no connections to the vertical input sections of the scope for this part.

(c) Adjust the controls listed in Table 1.3 and comment on the effects.

TABLE 1.3

Control	Observed Effect
Focus	
Intensity	
Y-position	
X-position	

Part 2 dc Voltage Measurements

(a) Set the dc/ac/Gnd switch to the Gnd position and adjust the Y-position control until the *zero-volt* reference is a line centered vertically on the screen.

(b) Once the zero-volt level is established, move the dc/ac/Gnd switch to the dc position and set the vertical sensitivity to 1 V/cm and connect one channel of the scope across the 1.5-V battery as shown in Fig. 1.5.

FIG. 1.5

Record the vertical shift below.

Vertical shift = _____ cm

Determine the voltage by multiplying by the vertical sensitivity. That is,

dc Voltage = (vertical shift)(vertical sensitivity)

= (_____)(_____)

= _____ V

Change the sensitivity to 0.5 V/cm and note the effect on the vertical shift. Recalculate the dc voltage with this new shift.

dc Voltage = (vertical shift)(vertical sensitivity)

= (_____)(_____)

= _____ V

How do the two measurements compare?

Which is more accurate? Why?

(c) Disconnect the 1.5-V battery and re-establish the zero-volt reference line. Then connect the vertical input section of the scope as shown in Fig. 1.6 with the vertical sensitivity set at 1 V/cm.

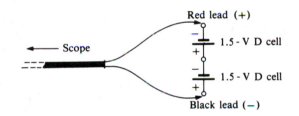

FIG. 1.6

What was the effect on the magnitude and direction of the shift?

Can a scope determine whether a particular voltage is positive or negative? How?

Calculate the measured voltage as follows:

dc Voltage = (vertical shift)(vertical sensitivity)

$$= (\text{_____})(\text{_____})$$

$$= \text{_____} V$$

Measure the voltage with the DMM (or VOM) and compare results.

dc Voltage [DMM (or VOM)] = _____ V

Part 3   Sinusoidal Waveforms—Magnitude

In this part of the experiment we will learn how to set the magnitude of a sinusoidal signal using an oscilloscope or DMM (or VOM). The frequency will remain fixed at 500 Hz.

Oscilloscope:

    (a) Connect the output of the oscillator or generator directly to one channel of
the scope as shown in Fig. 1.7.

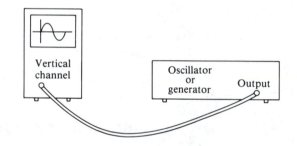

FIG. 1.7

    (b) Set the output frequency of the oscillator or generator to 500 Hz using the
dial and appropriate multiplier. Turn the amplitude knob all the way to the left for minimum output.

    (c) Set the vertical sensitivity of the scope to 1 V/cm and the horizontal sen-
sitivity to 0.5 ms/cm and turn on both the scope and the oscillator or generator.

    (d) Set the dc/ac/Gnd switch to the Gnd position to establish the zero-volt ref-
erence level (also the vertical center of a sinusoidal waveform) and then return the switch
to the ac position.

    (e) Now adjust the amplitude control of the oscillator or generator until the sig-
nal has a 4-V peak-to-peak swing. The resulting waveform has the following mathematical formulation:

$$v = V_m \sin 2\pi ft = 2 \sin 2\pi 500t$$

    (f) Switch to the dc position and comment below on any change in the
waveform.

    (g) Make the necessary adjustments to display the following waveforms on the
screen. Sketch both patterns in the space provided, showing the number of divisions (in

centimeters) for the vertical and horizontal distances, and the vertical and horizontal sensitivities.

1. $v = 0.2 \sin 2\pi 500t$

Vertical Sensitivity = _____

Horizontal Sensitivity = _____

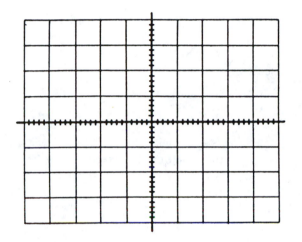

2. $v = 8 \sin 2\pi 500t$

Vertical Sensitivity = _____

Horizontal Sensitivity = _____

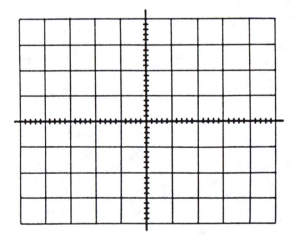

DMM (or VOM):

 (h) The sinusoidal signal $2 \sin 2\pi 500t$ has an effective value determined by

$$V_{eff} = 0.707 V_m = 0.707(2) = 1.414 \text{ V}$$

Connect the DMM (or VOM) directly across the oscillator in the ac rms mode and adjust the oscillator output until $V_{eff} = 1.414$ V. Then connect the output of the oscillator directly to the scope and note the total peak-to-peak swing.

Is the waveform the same as that obtained in part 3(e)?

(i) Use the DMM (or VOM) to set the following sinusoidal output from the oscillator:

$$v = 5 \sin 2\pi 500t$$

$V_{eff} = $ _____ V

Set V_{eff} with the DMM (or VOM) by adjusting the output of the oscillator, and place the signal on the screen.

Calculate the peak-to-peak voltage as follows:

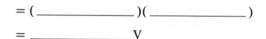

$V_{p-p} = $ (vertical distance peak to peak)(vertical sensitivity)

= (_____)(_____)

= _____ V

How does the above compare with the desired 10-V peak-to-peak voltage?

Part 4   Sinusoidal Waveforms – Frequency

This section will demonstrate how the oscilloscope can be used to set the frequency output of an oscillator or generator. In other words, the scope can be used to make fine adjustments on the frequency set by the dials of the oscillator or generator.

For a signal such as $2 \sin 2\pi 500t$, the frequency is 500 Hz and the period is $1/500 = 2$ ms. With a horizontal sensitivity of 0.5 ms/cm, the waveform should appear in exactly four horizontal divisions. If it does not, the fine-adjust control on the frequency of the oscillator or generator can be adjusted until it is exactly 4 cm. The scope has then set the output frequency of the oscillator.

Make the necessary adjustments to place the following waveforms on the scope. Sketch the waveforms in the space provided, indicating the number of vertical and horizontal deflections and the sensitivity of each.

1. $v = 0.4 \sin 60{,}280t$

$f = $ _____ Hz, $T = $ _____ s

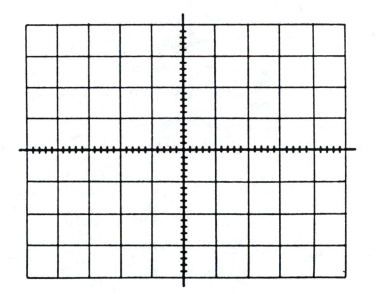

vertical deflection = _____ cm,

vertical sensitivity = _____

horizontal deflection = _____ cm,

horizontal sensitivity = _____

2. $v = 5 \sin 377t$

$f =$ _____ Hz, $T =$ _____ s

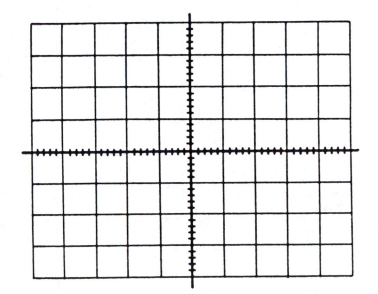

vertical deflection = _____ cm,

vertical sensitivity = _____

horizontal deflection = _____ cm,

horizontal sensitivity = _____

Part 5   Sinusoidal Waveforms on a dc Level

(a) Set the oscillator or generator to an output of $1 \sin 2\pi 500t$ using a vertical sensitivity of 1 V/cm on the scope with a horizontal sensitivity of 0.5 ms/cm.

(b) Measure the dc voltage of one of the D cells and insert in Fig. 1.8.

$E =$ _____ V

(c) Construct the network of Fig. 1.8.

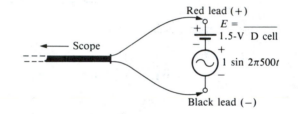

FIG. 1.8

(d) The input signal now has a dc level equal to the dc voltage of the D cell. Set the dc/ac/Gnd switch to the Gnd position and adjust the zero line to the center of the screen.

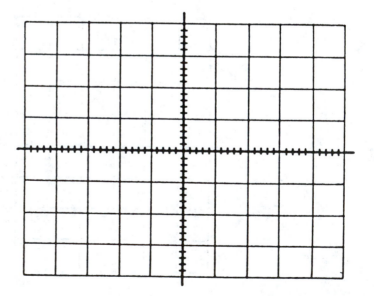

(e) Switch the ac mode and sketch the response, clearly indicating the zero-level line and the vertical deflection in centimetres.

(f) Switch to the dc mode and sketch the response, clearly indicating the zero-volt reference level and the vertical deflections in centimetres.

(g) How are the waveforms of parts 5(e) and 5(f) different? Why?

R-L-C Components

OBJECT

To measure the impedance of a resistor, inductor, and capacitor at a fixed frequency.

EQUIPMENT REQUIRED

Resistors

1 — 10-Ω, 100-Ω, 1.2-kΩ, 3.3-kΩ

Inductors

2 — 10-mH

Capacitors

1 — 0.5-μF, 1-μF

Instruments

1 — DMM (or VOM)

1 — Oscilloscope

1 — Audio oscillator (or signal generator)

EQUIPMENT ISSUED

TABLE 2.1

Item	Manufacturer and Model No.	Laboratory Serial No.
DMM (or VOM)		
Oscilloscope		
Audio oscillator (or signal generator)		

TABLE 2.2

Resistors	
Nominal Value	**Measured Value**
10 Ω	
100 Ω	
1.2 kΩ	
3.3 kΩ	

RÉSUMÉ OF THEORY

For impedances in series, the total impedance is the sum of the individual impedances:

$$\boxed{\mathbf{Z}_T = \mathbf{Z}_1 + \mathbf{Z}_2 + \mathbf{Z}_3 + \cdots + \mathbf{Z}_N} \tag{2.1}$$

For resistors in series,

$$\boxed{R_T = R_1 + R_2 + R_3 + \cdots + R_N} \tag{2.2}$$

which is independent of frequency.

For inductors in series,

$$X_{L_T} = X_1 + X_2 + X_3 + \cdots + X_N$$
$$= 2\pi f L_1 + 2\pi f L_2 + 2\pi f L_3 + \cdots + 2\pi f L_N$$
$$= 2\pi f L_T$$

and

$$\boxed{L_T = L_1 + L_2 + L_3 + \cdots + L_N} \tag{2.3}$$

Note that the individual and total inductive reactances are directly proportional to frequency.

For capacitors in series,

$$X_{C_T} = X_1 + X_2 + X_3 + \cdots + X_N$$

$$= \frac{1}{2\pi f C_1} + \frac{1}{2\pi f C_2} + \frac{1}{2\pi f C_3} + \cdots + \frac{1}{2\pi f C_N}$$

$$= \frac{1}{2\pi f C_T}$$

where

$$\boxed{\frac{1}{C_T} = \frac{1}{C_1} + \frac{1}{C_2} + \frac{1}{C_3} + \cdots + \frac{1}{C_N}} \tag{2.4}$$

Note that the individual and total capacitive reactances are inversely proportional to frequency.

Consider the special case of only two capacitors in series:

$$X_{C_T} = \frac{1}{2\pi f C_T}$$

where

$$C_T = \frac{C_1 C_2}{C_1 + C_2}$$

For resistors in parallel,

$$\boxed{\frac{1}{R_T} = \frac{1}{R_1} + \frac{1}{R_2} + \frac{1}{R_3} + \cdots + \frac{1}{R_N}} \tag{2.5}$$

which is independent of frequency.

For inductors in parallel,

$$\frac{1}{X_{L_T}} = \frac{1}{X_1} + \frac{1}{X_2} + \frac{1}{X_3} + \cdots + \frac{1}{X_N}$$

$$= \frac{1}{2\pi f L_1} + \frac{1}{2\pi f L_2} + \frac{1}{2\pi f L_3} + \cdots + \frac{1}{2\pi f L_N}$$

$$= \frac{1}{2\pi f L_T}$$

where

$$\boxed{\frac{1}{L_T} = \frac{1}{L_1} + \frac{1}{L_2} + \frac{1}{L_3} + \cdots + \frac{1}{L_N}} \tag{2.6}$$

Consider the special case of only two inductors in parallel:

$$X_{L_T} = \frac{X_1 X_2}{X_1 + X_2} = 2\pi f L_T$$

where

$$X_1 = 2\pi f L_1 \qquad X_2 = 2\pi f L_2 \qquad L_T = \frac{L_1 L_2}{L_1 + L_2}$$

For capacitors in parallel,

$$\frac{1}{X_{C_T}} = \frac{1}{X_1} + \frac{1}{X_2} + \frac{1}{X_3} + \cdots + \frac{1}{X_N}$$
$$= 2\pi f C_1 + 2\pi f C_2 + 2\pi f C_3 + \cdots + 2\pi f C_N$$
$$= 2\pi f C_T$$

where

$$\boxed{C_T = C_1 + C_2 + C_3 + \cdots + C_N} \qquad (2.7)$$

Consider the special case of only two capacitors in parallel:

$$X_{C_T} = \frac{X_1 X_2}{X_1 + X_2} = \frac{1}{2\pi f C_T}$$

where

$$C_T = C_1 + C_2$$

PROCEDURE

Part 1   Resistance

(a) Construct the circuit of Fig. 2.1. Insert the actual values of the resistors as determined by the ohmmeter section of your multimeter.

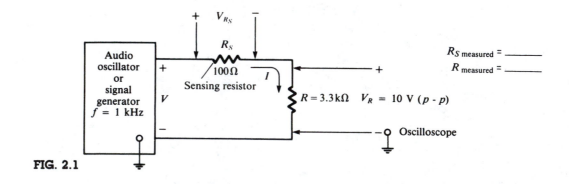

FIG. 2.1

Caution: Always ensure that the ground of the oscilloscope is connected to the ground of the oscillator. Otherwise a hazardous situation may result.

(b) Set the voltage across R to 10 V (p-p) using the oscilloscope. Measure the rms voltage across the sensing resistor (100 Ω) with the DMM (or VOM).

$V_{R_S} = $ _____

Calculate the peak value of V_{R_S}

$V_p = $ _____

Calculate the peak-to-peak value of V_{R_S}

$V_{p\text{-}p} = $ _____

(c) Calculate the peak-to-peak value of the current I from

$$I_{p\text{-}p} = \frac{V_{R_{S(p\text{-}p)}}}{R_S}$$

$I_{p\text{-}p} = $ _____

(d) Determine the resistance of the resistor R from

$$R = \frac{10 \text{ V } (p\text{-}p)}{I_{p\text{-}p}}$$

$R = $ _____

(e) Compare the value obtained in part 1(d) with the measured value inserted in Fig. 2.1.

(f) Connect a 1.2-kΩ and a 3.3-kΩ resistor in series as shown in Fig. 2.2

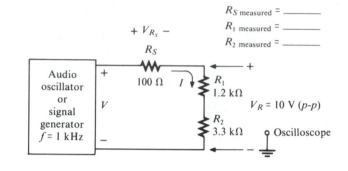

R_S measured = _____
R_1 measured = _____
R_2 measured = _____

FIG. 2.2

Calculate the total resistance of the series combination of the 1.2-kΩ and 3.3-kΩ resistors using the measured values.

$R_T = $ _____

(g) Set the voltage across the series combination to 10 V (*p-p*) and then measure the voltage V_{R_S} with the DMM or VOM.

$V_{R_S} = $ _____

(h) Calculate the peak-to-peak value of V_{R_S}.

$V_{R_{S(p-p)}} = $ _____

(i) Determine the peak-to-peak value of the current I from

$$I_{p-p} = \frac{V_{R_{S(p-p)}}}{R_S}$$

$I_{p-p} = $ _____

(j) Calculate the total resistance R_T of the series combination from

$$R_T = \frac{V_{R_{T(p-p)}}}{I_{p-p}} = \frac{10 \text{ V}}{I_{p-p}}$$

$$R_T = \underline{\hspace{2cm}}$$

(k) Compare the results of part 1(j) with the calculated value of part 1(f).

Part 2   Capacitive Reactance

(a) Construct the circuit of Fig. 2.3. Insert the measured value of R_S.

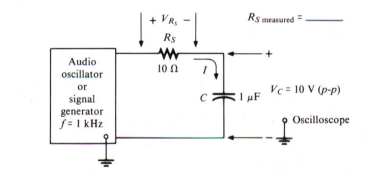

FIG. 2.3

(b) Set the voltage V_C to 10 V (*p-p*) and measure the rms value of the voltage V_{R_S} with the DMM (or VOM).

$$V_{R_S} = \underline{\hspace{2cm}}$$

(c) Using the technique described in part 1, calculate the peak-to-peak value of the current I.

$$I_{p\text{-}p} = \underline{\hspace{2cm}}$$

(d) Calculate the reactance X_C from the peak-to-peak values of V_C and I.

$$X_C = \frac{V_{C(p\text{-}p)}}{I_{p\text{-}p}}$$

X_C(measured) = _____

(e) Using the nameplate value of the capacitance, calculate the reactance of the capacitor at $f = 1$ kHz and compare with the results of part 2(d).

X_C(calculated) = _____

(f) Using the measured value of X_C in part 2(d), determine the capacitance level if $f = 1$ kHz and compare with the nameplate value of the capacitance.

$C = $ _____

(g) Connect a 0.5-μF in parallel with the capacitor of Fig. 2.3 and set V_C to 10 V (p-p) again. Measure the rms value of V_{R_S} using a DMM (or VOM).

$V_{R_S} = $ _____

Calculate the peak-to-peak value of V_{R_S}.

$V_{R_{S(p-p)}} = $ _____

Determine the peak-to-peak value of I from

$$I_{p-p} = \frac{V_{R_{S(p-p)}}}{R_S}$$

$I_{p-p} = $ _____

(h) Calculate the total reactance of the parallel capacitors from

$$X_{C_T} = \frac{V_{C(p\text{-}p)}}{I_{p\text{-}p}} = \frac{10 \text{ V}}{I_{p\text{-}p}}$$

$X_{C_T}(\text{measured}) = \underline{\hspace{3cm}}$

(i) Using the total capacitance as determined from the nameplate values, calculate the total reactance of the parallel capacitor at a frequency of $f = 1$ kHz.

$X_{C_T}(\text{calculated}) = \underline{\hspace{3cm}}$

Compare the result with the measured value of part 2(h).

(j) Using the measured value of X_{C_T} in part 2(h), calculate the equivalence total capacitance at $f = 1$ kHz and compare with the nameplate level.

$C_T = \underline{\hspace{3cm}}$

Part 3 Inductive Reactance

(a) Construct the network of Fig. 2.4. Insert the measured values for R_S and R_l.

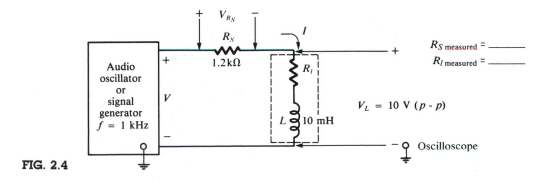

FIG. 2.4

(b) Set the voltage V_L to 10 V (*p-p*) and measure the rms value of the voltage V_{R_S} with a DMM (or VOM).

$$V_{R_S} = \text{\underline{\hspace{3cm}}}$$

(c) Calculate $V_{R_{S(p-p)}}$

$$V_{R_{S(p-p)}} = \text{\underline{\hspace{3cm}}}$$

(d) Calculate I_{p-p}

$$I_{p-p} = \text{\underline{\hspace{3cm}}}$$

(e) Ignoring the effects of R_1, calculate the reactance X_L from

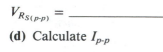

$$X_L = \frac{V_{L(p-p)}}{I_{p-p}} = \frac{10 \text{ V}}{I_{p-p}}$$

$$X_L \text{ (measured)} = \text{\underline{\hspace{3cm}}}$$

(f) Using the nameplate value of inductance L, calculate the inductive reactance at $f = 1$ kHz.

$$X_L \text{ (calculated)} = \text{\underline{\hspace{3cm}}}$$

(g) Compare the results of part 3(f) with the measured level of part 3(e).

(h) Calculate the inductance level necessary to establish an inductive reactance of the magnitude measured in part 3(e).

$$L = \text{_____}$$

(i) Compare the results of part 3(h) with the nameplate inductance level.

(j) Connect a second 10-mH coil in series with the coil of Fig. 2.4 and establish 10 V (*p-p*) across the two series coils. Then measure the rms value of V_{R_S} and calculate $V_{R_{S(p-p)}}$.

$$V_{R_{S(rms)}} = \text{_____}$$

$$V_{R_{S(p-p)}} = \text{_____}$$

(k) Calculate I_{p-p}.

$$I_{p-p} = \text{_____}$$

(l) Determine X_{L_T} from

$$X_{L_T} = \frac{V_{L(p-p)}}{I_{p-p}} = \frac{10 \text{ V}}{I_{p-p}}$$

$$X_{L_T}(\text{measured}) = \text{_____}$$

(m) Using the nameplate values of the inductances, calculate the total reactance at $f = 1$ kHz and compare with the results of part 3(l).

X_{L_T}(calculated) = _____

(n) Using the measured value of X_{L_T} in part 3(l), calculate the required inductance level at $f = 1$ kHz and compare with the total nameplate inductance level.

Frequency Response of *R* and *L* Components

OBJECT

To note the effect of frequency on the basic R and L components.

EQUIPMENT REQUIRED

Resistors

1—47-Ω, 100-Ω, 1-kΩ

Inductors

2—10-mH

Instruments

1—DMM
1—Oscilloscope
1—Audio oscillator (or function generator)

EQUIPMENT ISSUED

TABLE 3.1

Item	Manufacturer and Model No.	Laboratory Serial No.
DMM		
Oscilloscope		
Audio oscillator (or function generator)		

TABLE 3.2

Resistors	
Nominal Value	Measured Value
47 Ω	
100 Ω	
1 kΩ	

RÉSUMÉ OF THEORY

The resistance of a carbon resistor is unaffected by frequency, except for extremely high frequencies. This rule is also true for the total resistance of resistors in series or parallel.

The reactance of an inductor is linearly dependent on the frequency applied. That is, if we double the frequency, we double the reactance, as determined by $X_L = 2\pi f L$. For very low frequencies, the reactance is correspondingly very small, while for increasing frequencies, the reactance will increase to a very large value. For dc conditions, we find that $X_L = 2\pi(0)L$ is zero ohms, corresponding with the short-circuit representation we used in our analysis of dc circuits. For very high frequencies, X_L is so high that we can often use an open-circuit approximation.

PROCEDURE

Part 1   Resistors

Construct the circuit of Fig. 3.1. Insert the measured value of R.

In this part of the experiment, the voltage across the resistor will be held constant while only the frequency is varied. If the resistance is frequency independent, the current through the circuit should not change as a function of frequency. Therefore, by keeping the voltage V_{ab} constant and changing the frequency while monitoring the current I, we can verify if, indeed, resistance is frequency independent.

Set the voltage V_{ab} to 4 V (p-p) (measure this voltage with the oscilloscope). Then set the frequencies to those shown in Table 3.3, each time monitoring V_{ab} [always 4 V (p-p)] and I. Since we are using the DMM for current measurement, this value is given in rms. Record the values in Table 3.3.

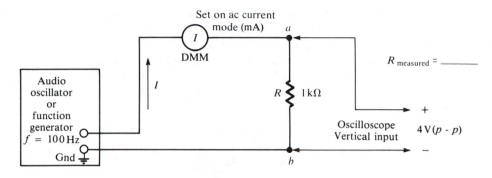

FIG. 3.1

TABLE 3.3

Frequency	$V_{ab(p-p)}$	$V_{ab(rms)} = \dfrac{0.707}{2} V_{ab(p-p)}$	I_{rms}	$R = \dfrac{V_{ab(rms)}}{I_{rms}}$
100 Hz	4 V	1.414 V		
200 Hz	4 V	1.414 V		
500 Hz	4 V	1.414 V		
1000 Hz	4 V	1.414 V		
2000 Hz	4 V	1.414 V		

Did the resistance *R* as calculated in the last column change with frequency?

Part 2   Inductors

(a) Construct the circuit of Fig. 3.2. (Measure the dc resistance of the coil with the DMM and record that value as R_l on the diagram.)

In this part, the resistor of part 1 is replaced by the inductor. Here again, the voltage across the inductor will be kept constant while we vary the frequency of that volt-

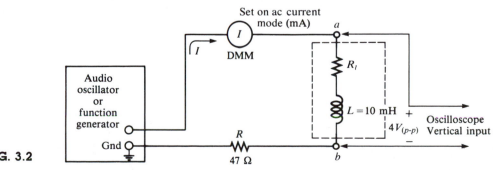

FIG. 3.2

age and monitor the current in the circuit. The 47-Ω resistor ensures that there is some resistance in the circuit for all frequency levels to limit the current.

Set the voltage V_{ab} to 4 V (p-p) (use the oscilloscope to measure V_{ab}) and the oscillator or generator to the various frequencies shown in Table 3.4, each time making sure that $V_{ab} = 4$ V (p-p) and that you measure the current I (remember I will be in rms). Record in Table 3.4.

TABLE 3.4

Frequency	$V_{ab(p\text{-}p)}$	$V_{ab(rms)}$	I_{rms}	$X_L \text{(measured)} = \dfrac{V_{ab(rms)}}{I_{rms}}$	$X_L \text{(calculated)} = 2\pi fL$
1 kHz	4 V	1.414 V			
3 kHz	4 V	1.414 V			
5 kHz	4 V	1.414 V			
7 kHz	4 V	1.414 V			
10 kHz	4 V	1.414 V			

(b) Calculate the reactance X_L (magnitude only) at each frequency and insert the values in Table 3.4 under the heading "X_L(measured)". We are assuming that $X_L \gg R_l$ for the frequencies of interest, so that

$$Z = \frac{V_{ab(rms)}}{I_{rms}} = \sqrt{R_l^2 + X_L^2} \cong \sqrt{X_L^2} = X_L$$

(c) Calculate the reactance at each frequency of Table 3.4 using the nameplate value of inductance (10 mH), and complete the table.

(d) Plot the measured and calculated values of X_L versus frequency on Graph 3.1. Label each curve and plot the points accurately.

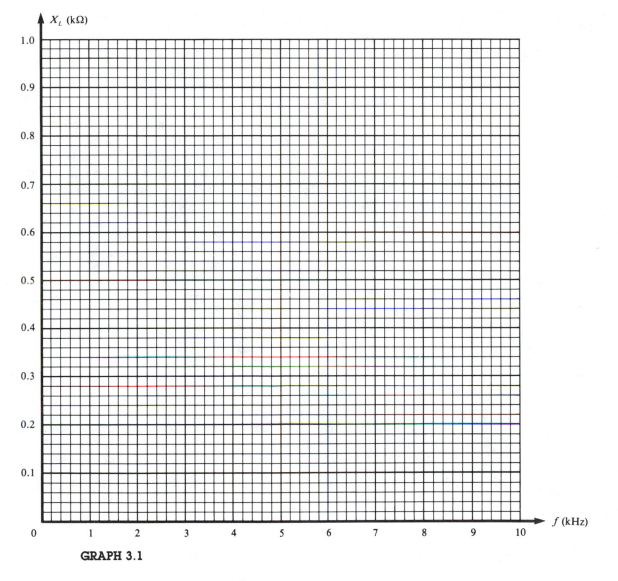

GRAPH 3.1

(e) Are both plots straight lines? Do they both pass through $X_L = 0$ at $f = 0$ Hz as they should?

(f) Determine the inductance at 1500 Hz using the results of part 2(d). That is, determine X_L from the graph and calculate L from $L = X_L/2\pi f$. Compare with the nameplate value.

$$L_{\text{calculated}} = \underline{\hspace{3cm}}, \quad L_{\text{nameplate}} = \underline{\hspace{3cm}}$$

Part 3   The Sensing Resistor

Whenever possible, the use of milliammeters is avoided in practice because the network has to be disturbed to insert the meter. In place of the meter is placed a sensing resistor that through Ohm's Law permits the calculation of the circuit current. In Fig. 3.3 a 100-Ω sensing resistor has been added in place of the DMM of Fig. 3.2. In addition, two coils are in parallel to change the net inductance level to $L_T = 10$ mH/2 $= 5$ mH. The effect of R_l is ignored in this part.

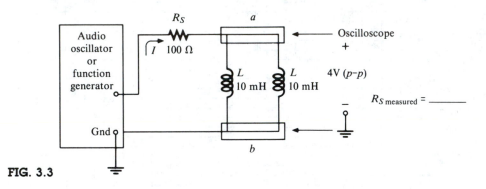

FIG. 3.3

(a) Construct the network of Fig. 3.3. With the generator set at 1 kHz, adjust the output of the supply until a 4-V peak-to-peak signal is obtained across the parallel coils. As noted in Part 2, the rms value of the voltage across the coils will be 1.414 V.

(b) For each frequency appearing in Table 3.5, set the voltage across the coils to 4 V (p-p) and measure the rms value of the voltage across the resistor R_S. Insert the values of $V_{R_S(\text{rms})}$ in Table 3.5.

TABLE 3.5

Frequency	$V_{ab(\text{rms})}$	$V_{R_S(\text{rms})}$	I_{rms}	$X_L\text{(measured)} = \dfrac{V_{ab(\text{rms})}}{I_{\text{rms}}}$	$X_L\text{(calculated)} = 2\pi fL$
1 kHz	1.414 V				
3 kHz	1.414 V				
5 kHz	1.414 V				
7 kHz	1.414 V				
10 kHz	1.414 V				

(c) For each frequency of Table 3.5, calculate the rms value of the current from

$$I_{rms} = \frac{V_{R_{S(rms)}}}{R_S(\text{measured})}$$

and insert in Table 3.5.

(d) Calculate X_L from the measured values at each frequency and insert in the X_L(measured) column of Table 3.5.

(e) Calculate X_L using the total nameplate inductance level of 5 mH and insert in the X_L(calculated) column.

(f) How do the results of the X_L(measured) column compare with the X_L(calculated) column?

(g) Plot X_L (measured) versus frequency on Graph 3.1. Is the graph a straight line? What was the impact of a reduced inductance level on the location of the graph?

(h) Determine X_L at a frequency of 6 kHz from the graph.

$X_L =$ _____

Calculate the equivalent inductance level at this frequency.

$L =$ _____

Frequency Response of Capacitive Components

OBJECT

To note the effect of frequency on the reactance of a capacitor.

EQUIPMENT REQUIRED

Resistors

$1-100\text{-}\Omega$

Capacitors

$1-0.1\text{-}\mu\text{F}, 0.5\text{-}\mu\text{F}$

Instruments

$1-$ DMM
$1-$ Oscilloscope
$1-$ Audio oscillator (or function generator)

EQUIPMENT ISSUED

TABLE 4.1

Item	Manufacturer and Model No.	Laboratory Serial No.
DMM		
Oscilloscope		
Audio oscillator (or function generator)		

TABLE 4.2

Resistors	
Nominal Value	Measured Value
100 Ω	

RÉSUMÉ OF THEORY

The effect of frequency on the impedance of a resistor and inductor was examined in Experiment 3. We will now investigate the change in the reactance of a capacitor with an increase in frequency. At low frequencies the reactance of a coil is quite low, whereas for a capacitor the reactance is quite high at low frequencies, often permitting the use of an open-circuit equivalent. At higher frequencies the reactance of a coil increases rapidly in a linear fashion, the reactance of a capacitor decreases in a nonlinear manner. In fact, it drops off more rapidly than the reactance of a coil increases. At very high frequencies the capacitor can be approximated by a short-circuit equivalency.

PROCEDURE

Part 1 X_C vs f

(a) Construct the circuit of Fig. 4.1.

The 100-Ω sensing resistor will be used to "sense" the current level in the network. With the generator set at 200 Hz, adjust the output of the supply until a 4-V

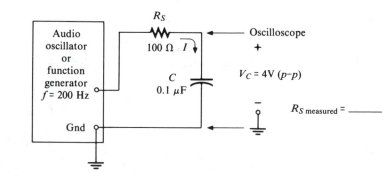

FIG. 4.1

peak-to-peak signal is obtained across the capacitor. The resulting rms voltage across the capacitor is then 1.414 V rms.

(b) For each frequency appearing in Table 4.3, set the voltage across the capacitor to 4 V (*p-p*) and measure the rms value of the voltage across the resistor R_S. Insert the values of $V_{R_{S(rms)}}$ in Table 4.3.

TABLE 4.3

Frequency	$V_{C(rms)}$	$V_{R_{S(rms)}}$	$I_{(rms)}$	$X_C(\text{measured}) = \dfrac{V_{C(rms)}}{I_{(rms)}}$	$X_C(\text{calculated}) = \dfrac{1}{2\pi fC}$
200 Hz	1.414 V				
500 Hz	1.414 V				
800 Hz	1.414 V				
1000 Hz	1.414 V				
2000 Hz	1.414 V				

(c) For each frequency in Table 4.3, calculate the rms value of the current from

$$I_{rms} = \frac{V_{R_{S(rms)}}}{R_S(\text{measured})}$$

and insert in Table 4.3.

(d) Calculate X_C from the measured values at each frequency and insert in the X_C(measured) column of Table 4.3.

(e) Calculate X_C using the nameplate capacitance level of 0.1 μF at each frequency and insert in the X_C(calculated) column.

(f) How do the results in the X_C(measured) column compare with those in the X_C(calculated) column?

(g) Plot X_C(measured) versus frequency on Graph 4.1. Is the graph linear or nonlinear? Where does the greatest change in capacitive reactance occur (in what frequency range)?

(h) Determine X_C at a frequency of 650 Hz from the curve just plotted.

$X_C = $ _____

At this frequency, determine the equivalent capacitance level using $X_C = 1/2\pi fC$

$C = $ _____

How does this level compare with the nameplate level of 0.1 μF?

Part 2 Parallel Capacitors (Increase in Capacitance)

(a) Place a 0.5-μF capacitor in parallel with the 0.1-μF capacitor of Fig. 4.1 to obtain a nameplate capacitance level of 0.6 μF ($C_T = C_1 + C_2$).

(b) Maintaining 4 V (p-p) across the parallel capacitors, measure and insert the rms value of V_{R_S} for each frequency in Table 4.4 using the technique described in part 1.

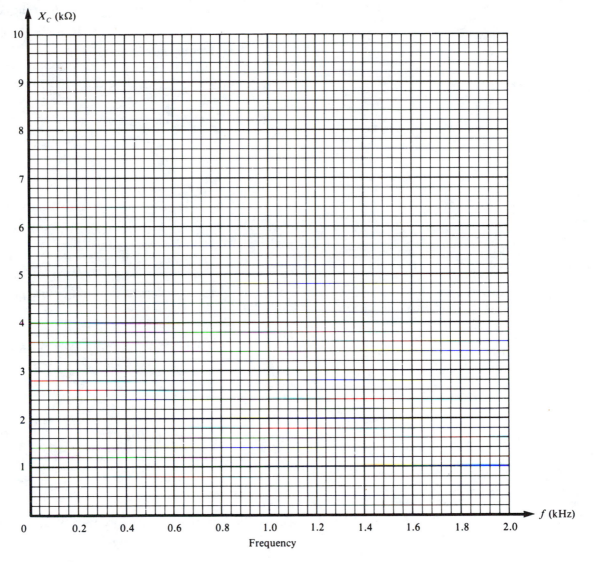

GRAPH 4.1

TABLE 4.4

Frequency	$V_{C(\text{rms})}$	$V_{R_{S(\text{rms})}}$	$I_{(\text{rms})}$	$X_C(\text{measured}) = \dfrac{V_{C(\text{rms})}}{I_{(\text{rms})}}$	$X_C(\text{calculated}) = \dfrac{1}{2\pi fC}$
200 Hz	1.414 V				
500 Hz	1.414 V				
800 Hz	1.414 V				
1000 Hz	1.414 V				
2000 Hz	1.414 V				

(c) For each frequency in Table 4.4, calculate the rms value of the current from

$$I_{rms} = \frac{V_{R_{S(rms)}}}{R_S(measured)}$$

and insert in Table 4.4.

(d) Calculate X_C from the measured values at each frequency and insert in the X_C(measured) column of Table 4.4.

(e) Calculate X_C using the nameplate capacitance level of 0.6 μF at each frequency and insert in Table 4.4 in the X_C(calculated) column.

(f) How do the results in the X_C(measured) column compare with those in the X_C(calculated) column?

(g) Plot X_C(measured) versus frequency on Graph 4.1. How did the change (increase) in capacitance level affect the location and characteristics of the curve?

(h) Determine X_C at a frequency of 1.5 kHz from the curve just plotted.

$X_C =$ _____

At this frequency, determine the equivalent capacitance level using $X_C = 1/(2\pi fC)$.

$C =$ _____

How does this capacitance level compare with the nameplate level of 0.6 μF?

Part 3 Series Capacitors (Reduced Capacitance Level)

(a) Place the 0.5-μF capacitor in series with the 0.1-μF capacitor of Fig. 4.1 to obtain a nameplate capacitance level of

$$C_T = \frac{(0.5 \ \mu\text{F})(0.1 \ \mu\text{F})}{0.5 \ \mu\text{F} + 0.1 \ \mu\text{F}} = 0.083 \ \mu\text{F}$$

(b) Maintaining 4 V (p-p) across the series combination of capacitors, measure and insert the rms value of V_{R_S} for each frequency in Table 4.5 using the technique described in part 1.

(c) For each frequency in Table 4.5, calculate the rms value of the current from

$$I_{\text{rms}} = \frac{V_{R_S(\text{rms})}}{R_S(\text{measured})}$$

and insert in Table 4.5.

TABLE 4.5

Frequency	$V_{C(\text{rms})}$	$V_{R_S(\text{rms})}$	$I_{(\text{rms})}$	$X_C(\text{measured}) = \dfrac{V_{C(\text{rms})}}{I_{(\text{rms})}}$	$X_C(\text{calculated}) = \dfrac{1}{2\pi fC}$
200 Hz					
500 Hz					
800 Hz					
1000 Hz					
2000 Hz					

(d) Calculate X_C from the measured values at each frequency and insert in the X_C(measured) column of Table 4.5.

(e) Calculate X_C using the nameplate capacitance level of 0.083 μF at each frequency and insert in Table 4.5 in the X_C(calculated) column.

(f) How do the results in the X_C(measured) column compare with those in the X_C(calculated) column?

(g) Plot X_C(measured) versus frequency in Graph 4.1. How did the change (decrease) in capacitance level affect the location and characteristics of the curve?

(h) Determine X_C at a frequency of 1.8 kHz from the curve just plotted.

$X_C =$ _____

At this frequency, determine the equivalent capacitance level using $X_C = 1/(2\pi fC)$.

$C =$ _____

How does this capacitance level compare with the nameplate level of 0.083 μF?

Frequency Response of the Series *R-L* Network

OBJECT

To plot the frequency response of the R-L series network.

EQUIPMENT REQUIRED

Resistors

1 — 100-Ω

Inductors

1 — 10-mH

Instruments

1 — DMM

1 — Oscilloscope

1 — Audio oscillator (or function generator)

EQUIPMENT ISSUED

TABLE 5.1

Item	Manufacturer and Model No.	Laboratory Serial No.
DMM		
Oscilloscope		
Audio oscillator (or function generator)		

TABLE 5.2

Resistors	
Nominal Value	**Measured Value**
100 Ω	
1 kΩ	

RÉSUMÉ OF THEORY

For the series dc or ac circuit, the voltage drop across a particular element is directly related to its impedance as compared with the other series elements. Since the impedances of the inductor and capacitor will change with frequency, the voltage across both elements will be determined by the applied frequency.

For the series R-L network, the voltage across the coil will increase with frequency since the inductive reactance increases directly with frequency and the impedance of the resistor is essentially independent of the applied frequency (in the audio range).

Since the voltage and current of the resistor are related by the fixed resistance value, the shapes of their curves versus frequency will be the same.

Keep in mind that the voltages across the elements in an ac circuit are vectorially related. Otherwise, the voltage readings may appear to be totally incorrect and not satisfy Kirchhoff's voltage law.

Caution: **Make sure that all measurements are made with the oscilloscope and oscillator sharing a common ground. The elements of the network may have to be reversed to ensure a common ground.**

PROCEDURE

Part 1 V_L, V_R, and I versus Frequency

(a) Construct the network of Fig. 5.1. Insert the measured values of the resistors R and R_l on the diagram. For the frequency range of interest, we will assume that $X_L \gg R_l$ and $\mathbf{Z}_L = \mathbf{X}_L$.

(b) Maintaining 4 V (p-p) at the input to the circuit, record the voltage $V_{L(p\text{-}p)}$ for the frequencies indicated in Table 5.3. Make sure to check continually that $E_S = 4$ V (p-p) with each frequency change.

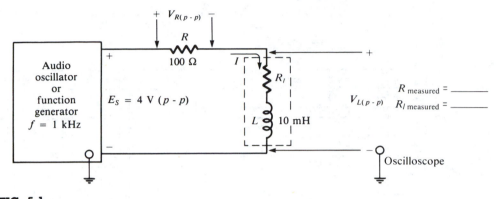

FIG. 5.1

TABLE 5.3

Frequency	$V_{L(p\text{-}p)}$	$V_{R(p\text{-}p)}$	$I_{(p\text{-}p)}$
1 kHz			
2 kHz			
3 kHz			
4 kHz			
5 kHz			
6 kHz			
7 kHz			
8 kHz			
9 kHz			
10 kHz			

    (c) Interchange the positions of R and L in Fig. 5.1 and measure $V_{R(p\text{-}p)}$ for the same range of frequencies with E_s maintained at 4 V (p-p). Insert the measurements in Table 5.3. **This is a very important step.** Failure to relocate the resistor R can result in a grounding situation where the inductive reactance is "shorted-out"!

    (d) Calculate $I_{p\text{-}p}$ from $I_{p\text{-}p} = V_{R(p\text{-}p)}/R_{\text{measured}}$ and complete Table 5.3. Use the space below for your calculations.

(e) Sketch the magnitudes of $V_{L(p\text{-}p)}$ and $V_{R(p\text{-}p)}$ versus frequency on Graph 5.1.

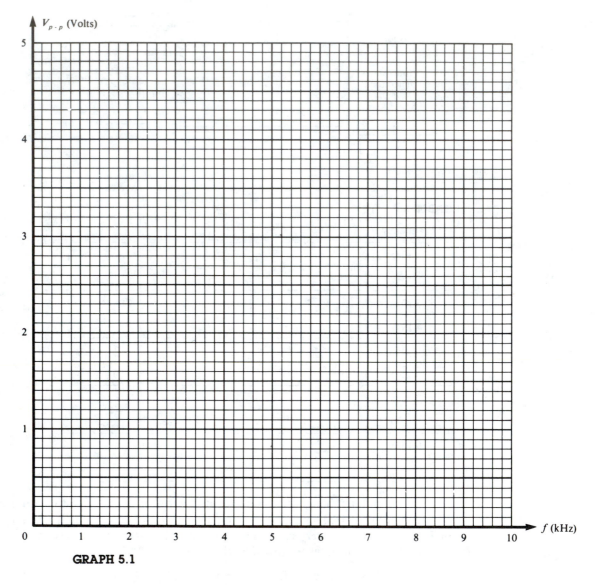

GRAPH 5.1

(f) At 5 kHz, does the magnitude of $V_{L(p\text{-}p)} + V_{R(p\text{-}p)} = E_{S(p\text{-}p)}$? Comment accordingly. How are they related?

(g) Sketch the magnitude of $I_{p\text{-}p}$ versus frequency on Graph 5.2. How does it compare with the graph of $V_{R(p\text{-}p)}$ versus frequency?

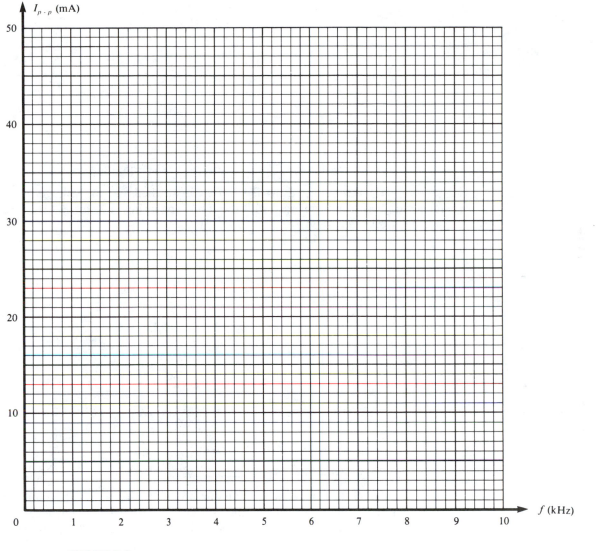

GRAPH 5.2

(h) At a frequency of 8 kHz, calculate the reactance of the inductor using $X_L = 2\pi fL$. Compare with the value obtained from the data of Table 5.3 using

$$X_L \cong \frac{V_{L(p\text{-}p)}}{I_{p\text{-}p}}$$

X_L(calculated) = _____ , X_L(from data) = _____

(i) Determine, using the Pythagorean theorem, the voltage $V_{L(p\text{-}p)}$ at a frequency of 5 kHz and compare with the result appearing in Table 5.3. Use peak-to-peak values for all calculations.

$V_{L(p\text{-}p)}$(calculated) = _____ ,

$V_{L(p\text{-}p)}$(measured) = _____

(j) At low frequencies the inductor approaches a low-impedance short-circuit equivalent and at high frequencies a high-impedance open-circuit equivalent. Does the data of Table 5.3 and Graphs 5.1 and 5.2 verify the above statement? Comment accordingly.

Part 2 Z_T versus Frequency

(a) Transfer the results of $I(p\text{-}p)$ from Table 5.3 to Table 5.4 for each frequency.

TABLE 5.4

Frequency	$E_{S(p\text{-}p)}$	$I_{(p\text{-}p)}$	$Z_T = \dfrac{E_{S(p\text{-}p)}}{I_{(p\text{-}p)}}$	$Z_T = \sqrt{R^2 + X_L^2}$
1 kHz	4 V			
2 kHz	4 V			
3 kHz	4 V			
4 kHz	4 V			
5 kHz	4 V			
6 kHz	4 V			
7 kHz	4 V			
8 kHz	4 V			
9 kHz	4 V			
10 kHz	4 V			

(b) At each frequency, calculate the magnitude of the total impedance using the equation $Z_T = E_{S(p\text{-}p)}/I_{(p\text{-}p)}$ in Table 5.4. Use the space below for your calculations.

(c) Plot the magnitude of the impedance $\mathbf{Z}_T$ versus frequency on Graph 5.3.

(d) For each frequency calculate the total impedance using the equation $Z_T = \sqrt{R^2 + X_L^2}$ and insert in Table 5.4. Use the space below for your calculations.

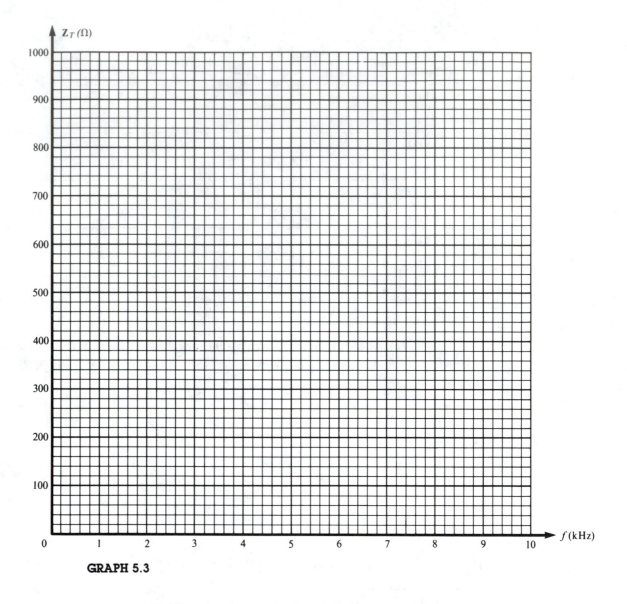

GRAPH 5.3

(e) How do the magnitudes of Z_T compare for the last two columns of Table 5.4?

(f) On the same graph (5.3), plot R versus frequency.

(g) On Graph 5.3, plot $X_L = 2\pi fL$ versus frequency. Use the space below for the necessary calculations.

(**h**) At which frequency does $X_L = R$? Use both the graph and a calculation $(f = R/2\pi L)$.

$f =$ _____ (graph), $f =$ _____ (calculation)

(**i**) For frequencies less than the frequency calculated in part 2(h) above, is the network primarily resistive or inductive? How about for frequencies greater than the frequency calculated in part 2(h)?

(**j**) The phase angle by which the applied voltage leads the same current is determined by $\theta = \tan^{-1}(X_L/R)$ (as obtained from the impedance diagram). Determine the phase angle for each of the frequencies in Table 5.5.

TABLE 5.5

Frequency	R (measured)	X_L	$\theta = \tan^{-1}(X_L/R)$
0.5 kHz			
1 kHz			
2 kHz			
3 kHz			
5 kHz			
10 kHz			

(k) At a frequency of 0.5 kHz, does the phase angle suggest a primarily resistive or inductive network? Explain why.

(l) At frequencies greater than 2 kHz, does the phase angle suggest a primarily resistive or inductive network? Explain why.

Frequency Response of the Series *R-C* Network

OBJECT

To plot the frequency response of the R-C series network.

EQUIPMENT REQUIRED

Resistors

1 — 1-kΩ

Capacitors

1 — 0.1-μF

Instruments

1 — DMM

1 — Oscilloscope

1 — Audio oscillator (or function generator)

EQUIPMENT ISSUED

TABLE 6.1

Item	Manufacturer and Model No.	Laboratory Serial No.
DMM		
Oscilloscope		
Audio oscillator (or function generator)		

TABLE 6.2

Resistors	
Nominal Value	Measured Value
1 kΩ	

RÉSUMÉ OF THEORY

As noted in Experiment 5 for the series R-L network, the voltage across the coil increases with frequency since the inductive reactance increases directly with frequency and the impedance of the resistor is essentially independent of the applied frequency (in the audio range).

For the series R-C network, the voltage across the capacitor decreases with increasing frequency since the capacitive reactance is inversely proportional to the applied frequency.

Since the voltage and current of the resistor continue to be related by the fixed resistance value, the shapes of their curves versus frequency will be the same.

Again, keep in mind that the voltages across the elements in an ac circuit are vectorially related. Otherwise, the voltage readings may appear to be totally incorrect and not satisfy Kirchhoff's voltage law.

Caution: **Make sure that all measurements are made with the oscilloscope and oscillator sharing a common ground. The elements of the network may have to be reversed to ensure a common ground.**

Part 1 V_C, V_R, and I versus Frequency

      (a) Construct the network of Fig. 6.1. Insert the measured value of the resistor R.

      (b) Maintaining 4 V (p-p) at the input to the circuit, record the voltage $V_{C(p\text{-}p)}$ for the frequencies indicated in Table 6.3. Make sure to check continually that $E_S = 4$ V (p-p) with each frequency change.

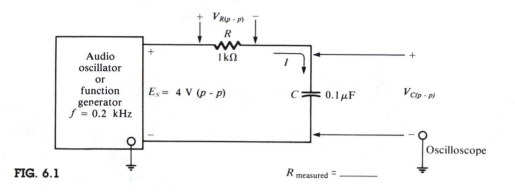

FIG. 6.1

(c) Interchange the positions of R and C in Fig. 6.1 and measure $V_{R(p-p)}$ for the same range of frequencies with E_S maintained at 4 V (p-p). Insert the measurements in Table 6.3. As in Experiment 5, it is vitally important that this step be performed as specified, or a grounding problem can result.

(d) Calculate I_{p-p} from $I_{p-p} = V_{R(p-p)}/R_{measured}$ and complete Table 6.3. Use the space below for your calculations.

TABLE 6.3

Frequency	$V_{C(p-p)}$	$V_{R(p-p)}$	$I_{(p-p)}$
0.2 kHz			
0.5 kHz			
1 kHz			
2 kHz			
4 kHz			
6 kHz			
8 kHz			
10 kHz			

(e) Sketch the magnitudes of $V_{C(p\text{-}p)}$ and $V_{R(p\text{-}p)}$ versus frequency on Graph 6.1.

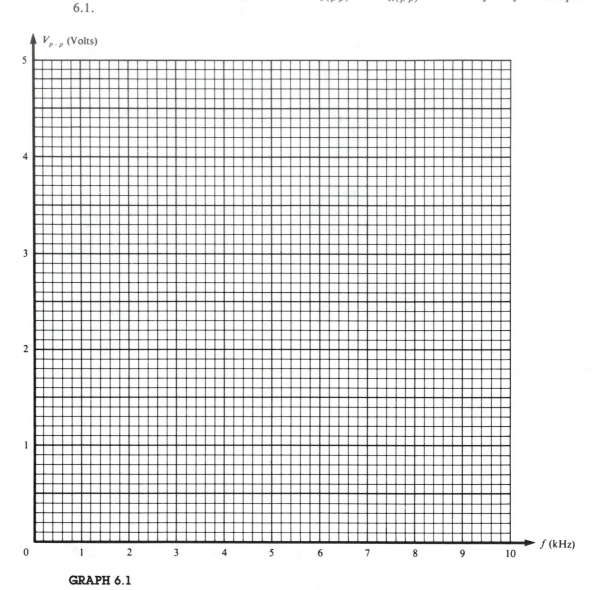

GRAPH 6.1

(f) At 10 kHz, does the magnitude of $V_{L(p\text{-}p)} + V_{R(p\text{-}p)} = E_{S(p\text{-}p)}$? Comment accordingly. How are they related?

(g) Sketch the magnitude of $I_{p\text{-}p}$ versus frequency on Graph 6.2. How does it compare with the graph of $V_{R(p\text{-}p)}$ versus frequency?

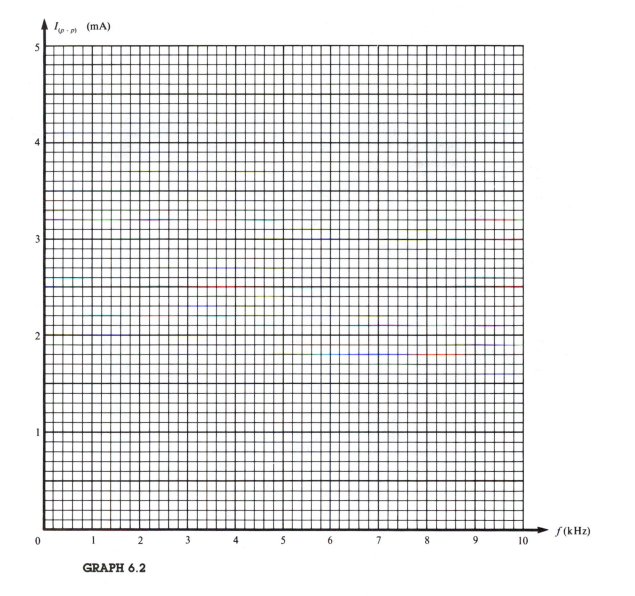

GRAPH 6.2

(h) At a frequency of 4 kHz, calculate the reactance of the capacitor using $X_C = 1/2\pi fC$. Compare with the value obtained from the data of Table 6.3 using

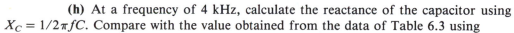

$$X_C = \frac{V_{C(p-p)}}{I_{p-p}}$$

X_C(calculated) = _____ , X_C(from data) = _____

(i) Determine, using the Pythagorean theorem, the voltage $V_{C(p-p)}$ at a frequency of 6 kHz and compare with the result in Table 6.3. Use peak-to-peak values for all calculations.

$V_{C(p-p)}$ (calculated) = _____ ,

$V_{C(p-p)}$ (measured) = _____

(j) At low frequencies the capacitor approaches a high-impedance open-circuit equivalent and at high frequencies a low-impedance short-circuit equivalent. Does the data of Table 6.3 and Graphs 6.1 and 6.2 verify the above statement? Comment accordingly.

Part 2 Z_T versus Frequency

(a) Transfer the results of $I_{(p-p)}$ from Table 6.3 to Table 6.4 for each frequency.

TABLE 6.4

Frequency	$E_{S(p\text{-}p)}$	$I_{(p\text{-}p)}$	$Z_T = \dfrac{E_{S(p\text{-}p)}}{I_{(p\text{-}p)}}$	$Z_T = \sqrt{R^2 + X_C^2}$
0.2 kHz	4			
0.5 kHz	4			
1 kHz	4			
2 kHz	4			
4 kHz	4			
6 kHz	4			
8 kHz	4			
10 kHz	4			

(b) At each frequency, calculate the magnitude of the total impedance using the equation $Z_T = E_{S(p\text{-}p)}/I_{(p\text{-}p)}$ in Table 6.4. Use the space below for your calculations.

(c) Plot the magnitude of the impedance Z_T versus frequency on Graph 6.3.

(d) For each frequency calculate the total impedance using the equation $Z_T = \sqrt{R^2 + X_C^2}$ and insert in Table 6.4. Use the space below for your calculations.

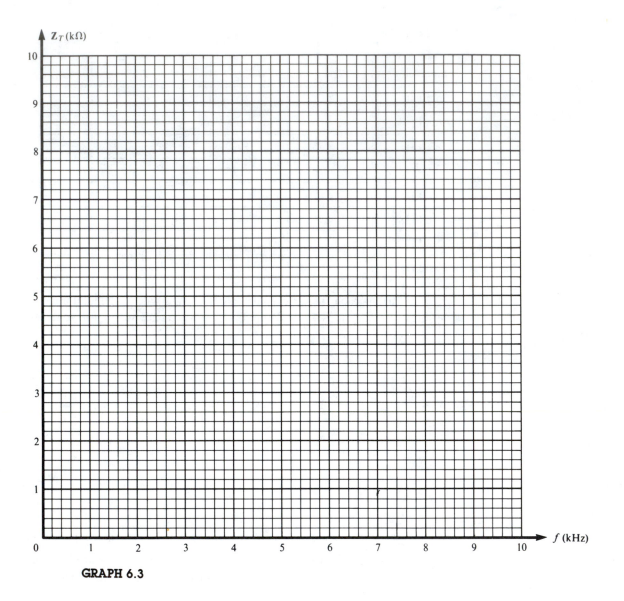

GRAPH 6.3

(e) How do the magnitudes of Z_T compare for the last two columns of Table 6.4?

(f) On Graph 6.3, plot R versus frequency.

(g) On Graph 6.3, plot $X_C = 1/2\pi fC$ versus frequency. Use the space below for your calculations.

(h) At which frequency does $X_C = R$? Use both the graph and a calculation $(f = 1/2\pi RC)$.

$f =$ _____ (graph), $f =$ _____ (calculation)

(i) For frequencies less than the frequency calculated in part 2(h) above, is the network primarily resistive or inductive? How about for frequencies greater than the frequency calculated in part 2(h)?

(j) The phase angle by which the source current leads the applied voltage is determined by $\theta = \tan^{-1}(X_C/R)$ (as obtained from the impedance diagram). Determine the phase angle for each of the frequencies in Table 6.5.

TABLE 6.5

Frequency	R (measured)	X_C	$\theta = \tan^{-1}(X_C/R)$
0.2 kHz			
0.5 kHz			
1 kHz			
2 kHz			
6 kHz			
10 kHz			

(k) At a frequency of 0.2 kHz, does the phase angle suggest a primarily resistive or capacitive network? Explain why.

(l) At frequencies greater than 2 kHz, does the phase angle suggest a primarily resistive or capacitive network? Explain why.

The Oscilloscope and Phase Measurements

OBJECT

*To use the oscilloscope to determine the phase
angle between two sinusoidal waveforms.*

EQUIPMENT REQUIRED

Resistors

1 — 1-kΩ
1 — 0–10-kΩ potentiometer

Capacitors

1 — 0.5-μF

Instruments

1 — DMM
1 — Oscilloscope
1 — Audio oscillator or function generator

EQUIPMENT ISSUED

TABLE 7.1

Item	Manufacturer and Model No.	Laboratory Serial No.
DMM		
Oscilloscope		
Audio oscillator or function generator		

TABLE 7.2

Resistors	
Nominal Value	Measured Value
1 kΩ	

PHASE MEASUREMENTS

The phase angle between two signals of the same frequency can be determined using the oscilloscope. There are two methods available.

1. Dual-trace comparison with the calibrated time base
2. Lissajous pattern

Dual-Trace Method of Phase Measurement

The dual-trace method of phase measurement, aside from providing a high degree of accuracy, can compare two signals of different amplitudes and, in fact, different wave shapes. The method can be applied directly to oscilloscopes equipped with two vertical channels or to a conventional single-trace oscilloscope with an external electronic switch, as shown in Fig. 7.1. The electronic switch will switch between inputs at a very high speed, so both patterns will appear on the screen.

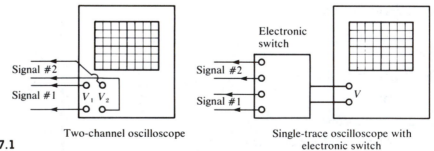

Signal #2
Signal #1
V_1 V_2

Two-channel oscilloscope

Electronic switch
Signal #2
Signal #1
V

Single-trace oscilloscope with electronic switch

FIG. 7.1

Regardless of which oscilloscope is available, the procedure essentially consists of displaying both traces on the screen simultaneously, and measuring the distance (in scale divisions) between two identical points on the two traces (Fig. 7.2).

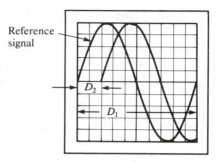

Reference signal

FIG. 7.2

One signal will be chosen as a reference, that is, zero-phase angle. In the comparison, therefore, we can assume that the signal being compared is leading ($+\theta$) if it is to the left of the reference and lagging ($-\theta$) if it is to the right of the reference. To use the dual-trace phase measurement method, therefore, proceed as follows:

1. Connect the two signals to the two vertical channels, making sure to observe proper grounding.
2. Select the mode of operation—"Alternate" or "Chop." For frequencies less than 50 kHz, use "Chop." For frequencies greater than 50 kHz, use "Alternate."
3. Once the traces are on the screen, use the Gnd switch to set both patterns in the vertical center of the screen.
4. Measure the number of horizontal divisions (in centimetres) required for one full cycle of either waveform (D_1 in Fig. 7.2).
5. Determine the scale factor SF from

$$SF = \frac{360°}{D_1} \qquad \text{(degrees/division)} \qquad \text{(7.1)}$$

6. Determine the number of horizontal divisions (in centimetres) between corresponding positive slopes of the waveforms, that is, the phase shift between the waveforms (D_2 in Fig. 7.2).
7. Calculate the phase angle from

$$\theta = (SF)(D_2) \qquad \text{(7.2)}$$

or

$$\theta = \left(\frac{D_2}{D_1}\right) 360° \qquad \text{(7.3)}$$

Lissajous-Pattern Phase Measurement

The Lissajous-pattern method is also called the X-Y phase measurement. To use this method, proceed as follows:

1. Connect the two signals to the vertical and the horizontal inputs, respectively.
2. A pattern known as a *Lissajous* will appear on the screen. The type of pattern will indicate certain common phase relationships, and in fact the patterns can be used to calculate the phase angle in general.

The patterns shown in Fig. 7.3 indicate the phase relationship appearing with each figure.

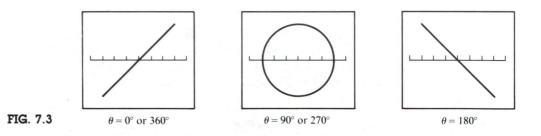

FIG. 7.3 $\theta = 0°$ or $360°$ $\theta = 90°$ or $270°$ $\theta = 180°$

The patterns shown in Fig. 7.4 can be used to calculate the phase angle (θ) as indicated below the figures.

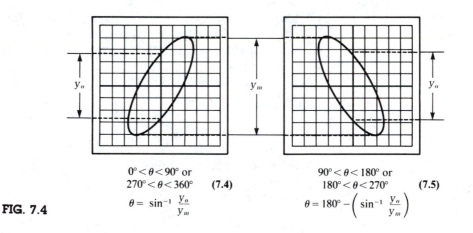

$$0° < \theta < 90° \text{ or} \qquad\qquad 90° < \theta < 180° \text{ or}$$
$$270° < \theta < 360° \quad \textbf{(7.4)} \qquad 180° < \theta < 270° \quad \textbf{(7.5)}$$

FIG. 7.4 $\qquad\qquad \theta = \sin^{-1} \dfrac{y_o}{y_m} \qquad\qquad \theta = 180° - \left(\sin^{-1} \dfrac{y_o}{y_m} \right)$

EXAMPLE   Assume that the patterns in Figs. 7.5 and 7.6 appear on an oscilloscope screen. Calculate the phase angle θ in each case.

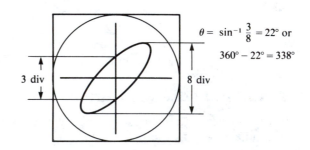

$\theta = \sin^{-1} \dfrac{3}{8} = 22°$ or

$360° - 22° = 338°$

3 div   8 div

FIG. 7.5

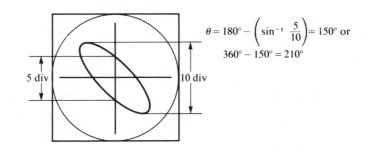

$$\theta = 180° - \left(\sin^{-1}\frac{5}{10}\right) = 150° \text{ or}$$

$$360° - 150° = 210°$$

FIG. 7.6

PROCEDURE

Part 1

The phase relationship between **E** and $\mathbf{V}_R$ of Fig. 7.7 will be determined in this part using the dual-trace capability of the oscilloscope.

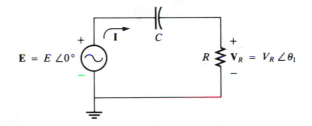

FIG. 7.7

Since **E** is defined as having an angle of zero degrees, it will appear as shown in Fig. 7.8. In an *R-C* circuit, the current **I** will lead the applied voltage as shown in the phasor diagram. The voltage $\mathbf{V}_R$ is in phase with **I**, and the voltage $\mathbf{V}_C$ will lag the voltage **E**.

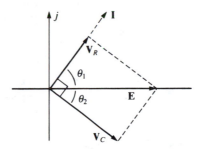

FIG. 7.8

Note that $\theta_1 + \theta_2 = 90°$ and that the vector sum of $\mathbf{V}_R$ and $\mathbf{V}_C$ equals the applied voltage **E**.

Construct the network of Fig. 7.9, designed to place the voltage $\mathbf{V}_R$ across channel #1 of the scope and **E** across channel #2. Note that the resistance R is composed of a series combination of the fixed 1-kΩ resistor and 10-kΩ potentiometer. That is, $R = 1 \text{ k}\Omega + 10 \text{ k}\Omega$.

The oscillator or function generator is set to 200 Hz with an amplitude of 8 V (*p-p*). Set the amplitude and frequency using channel #2 of the oscilloscope.

Set the potentiometer to zero ohms and calculate the reactance of the capacitor at a frequency of 200 Hz and insert in each row of Table 7.3.

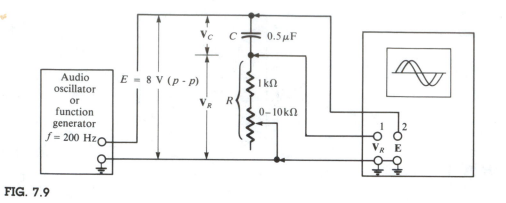

FIG. 7.9

TABLE 7.3

Potentiometer Setting	R	X_C	$E_{\text{rms}} = 0.707V_p$	$V_{R(\text{rms})}$ Calculated	$V_{R(\text{rms})}$ Measured	θ_1 Calculated	θ_1 Measured
0 Ω	1000 Ω		2.828 V				
2000 Ω	3000 Ω		2.828 V				
4000 Ω	5000 Ω		2.828 V				

(a) Using $R = 1000\ \Omega$, the calculated value of X_C, and $\mathbf{E} = 2.828\ \angle 0°$, calculate the rms value of $\mathbf{V}_R$ and the angle θ_1 associated with the resistance as defined in Fig. 7.8 and insert in Table 7.3.

In general,

$$|V_R| = \frac{RE}{\sqrt{R^2 + X_C^2}} \qquad (7.6)$$

with

$$\theta_1 = \tan^{-1} \frac{X_C}{R} \qquad (7.7)$$

$R = 1000\text{-}\Omega$ **Calculations:**

Repeat the above for $R = 3000$ Ω and 5000 Ω and insert the calculated results in Table 7.3.

$R = 3000$-Ω Calculations:

$R = 5000$-Ω Calculations:

(b)

$R = 1000$-Ω Measurements:

Measure $V_{R(p\text{-}p)}$ from the scope display and calculate $V_{R(\text{rms})}$ and insert in Table 7.3 for $R = 1000$ Ω.

$V_{R(p\text{-}p)} =$ _____ , $V_{R(\text{rms})} =$ _____

Determine the number of horizontal divisions for one full cycle of **E** or **V_R**.

Number of horizontal divisions (in centimetres) for one full cycle =

$D_1 =$ _____

Determine the number of horizontal divisions representing the phase shift between waveforms.

Number of horizontal divisions (in centimetres) between waveforms =

$D_2 =$ _____

Determine the phase shift in degrees from

$$\boxed{\theta_1 = \left(\frac{D_2}{D_1}\right)360°}$$ (7.8)

$\theta_1 = ($ _____ $) 360° = $ _____ °

Insert the measured value of θ_1 in Table 7.3.

Repeat the above for $R = 3000\ \Omega$ and $5000\ \Omega$ and insert the results in Table 7.3.

$R = 3000\text{-}\Omega$ Measurements:

$V_{R(p\text{-}p)} = $ _____ , $V_{R(rms)} = $ _____

$D_1 = $ _____ cm, $D_2 = $ _____ cm

$\theta_1 = $ _____ °

$R = 5000\text{-}\Omega$ Measurements:

$V_{R(p\text{-}p)} = $ _____ , $V_{R(rms)} = $ _____

$D_1 = $ _____ cm, $D_2 = $ _____ cm

$\theta_1 = $ _____ °

(c) Calculate the percent difference between the calculated and measured values of θ_1 from Table 7.3 using the following equation, and insert in Table 7.4.

$$\boxed{\% \text{ Difference } (\theta_1) = \frac{|\text{calculated} - \text{measured}|}{\text{calculated}} \times 100\%}$$ (7.9)

TABLE 7.4

R	% Difference in θ_1
1000 Ω	
3000 Ω	
5000 Ω	

(d) The vector $\mathbf{E} = E \angle \theta = 2.828$ V $\angle 0°$ has been placed on each phasor diagram of Graph 7.1. Note that the voltage has been scaled to match the 1 V/cm scale of the horizontal and vertical axes. Using the measured values of $V_{R(\text{rms})}$ and θ_1, insert the phasor $\mathbf{V}_R$ for each value of R. Clearly indicate the angle θ_1 and the magnitude of $V_{R(\text{rms})}$.

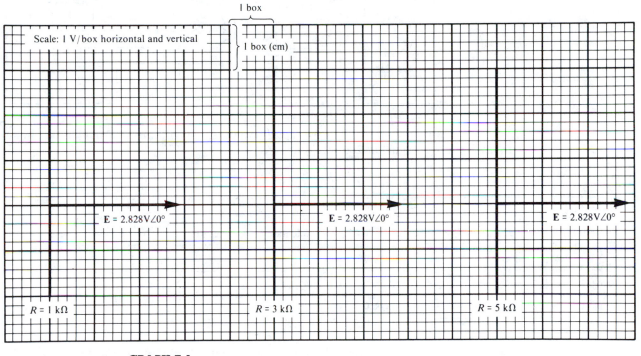

GRAPH 7.1

Part 2

The phase relationship between $\mathbf{E}$ and $\mathbf{V}_C$ of the same network as Fig. 7.7 will now be determined by interchanging the resistor R and capacitor C, as shown in Fig. 7.10.

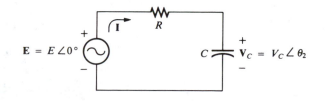

FIG. 7.10

The elements must exchange positions to avoid a "shorting out" of the resistor *R* of Fig. 7.7 if the oscilloscope were simply placed across the capacitor of Fig. 7.7. The grounds of the scope and supply would establish a zero-volt drop across *R* and possibly high currents in the remaining network since X_C is the only impedance to limit the current level.

The connections are now made as shown in Fig. 7.11. Note that **E** remains on channel #2 and $\mathbf{V}_C$ is placed on channel #1. The phase angle θ_2 between **E** and $\mathbf{V}_C$ can therefore be determined from the display.

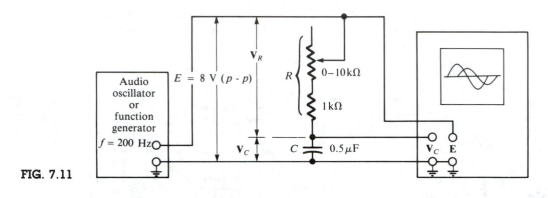

FIG. 7.11

Insert the reactance of the capacitor on each line of Table 7.5 and set the supply to an 8 V (*p-p*) signal at 200 Hz.

TABLE 7.5

Potentiometer Setting	R	X_C	$E_{rms} = 0.707 V_p$	$V_{C(rms)}$		θ_2	
				Calculated	Measured	Calculated	Measured
0 Ω	1000 Ω		2.828 V				
2000 Ω	3000 Ω		2.828 V				
4000 Ω	5000 Ω		2.828 V				

(a) Using $R = 1000\ \Omega$, the calculated value of X_C, and $E = 2.828$ V $\angle 0°$, calculate the rms value of V_C and the angle θ_2 associated with the reactance as defined by Fig. 7.8 and insert in Table 7.5.

In general,

$$|V_C| = \frac{X_C E}{\sqrt{R^2 + X_C^2}} \qquad\qquad (7.10)$$

with

$$\theta_2 = -90° + \tan^{-1} \frac{X_C}{R} \qquad\qquad (7.11)$$

$R = 1000$-Ω **Calculations:**

Repeat the above for $R = 3000\ \Omega$ and $5000\ \Omega$ and insert the calculated results in Table 7.5.

$R = 3000$-Ω **Calculations:**

$R = 5000$-Ω **Calculations:**

(b)

$R = 1000$-Ω **Measurements:**

Measure $V_{C(p\text{-}p)}$ from the scope display and calculate $V_{C(\text{rms})}$ and insert in Table 7.5 for $R = 1000\ \Omega$.

$V_{C(p\text{-}p)} =$ _____ , $V_{C(rms)} =$ _____

Determine the phase shift in the same manner as described for part 1.

$D_1 =$ _____ cm, $D_2 =$ _____ cm

$\theta_2 =$ _____ °

Repeat the above for $R = 3000\ \Omega$ and $5000\ \Omega$ and insert the results in Table 7.5.

$R = 3000\text{-}\Omega$ Measurements:

$V_{C(p\text{-}p)} =$ _____ , $V_{C(rms)} =$ _____

$D_1 =$ _____ cm, $D_2 =$ _____ cm

$\theta_2 =$ _____ °

$R = 5000\text{-}\Omega$ Measurements:

$V_{C(p\text{-}p)} =$ _____ , $V_{C(rms)} =$ _____

$D_1 =$ _____ cm, $D_2 =$ _____ cm

$\theta_2 =$ _____ °

(c) Complete the phasor diagrams of Graph 7.1 with the insertion of the phasor V_C for each resistance level.

(d) It was noted early in part 1 that $\theta_1 + \theta_2 = 90°$. For each resistance level, add the two and determine the percent difference using the equation

$$\% \text{ Difference} = \frac{|90° - \theta_T|}{90°} \times 100\% \tag{7.12}$$

Record the values in Table 7.6.

TABLE 7.6

R	θ_1 (part 1)	θ_2 (part 2)	$\theta_T = \theta_1 + \theta_2$	% **Difference** 90° vs. θ_T
1000 Ω				
3000 Ω				
5000 Ω				

Part 3

In this part, the phase angle will be determined from the Lissajous pattern.

Construct the network of Fig. 7.12, which is exactly the same as Fig. 7.9 except now the applied voltage **E** is placed on channel #1 with V_R applied as a *horizontal input*.

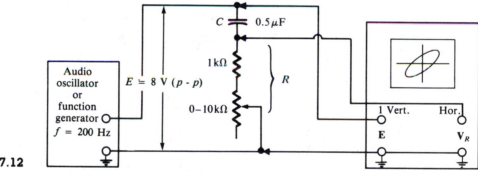

FIG. 7.12

The resulting Lissajous patterns will determine the phase angle between **E** and V_R for various values of R.

Set the value of R as indicated in Table 7.7 and measure the y-intercept (y_o) and the y-maximum (y_m) from the Lissajous pattern. Record in Table 7.7.

TABLE 7.7

Potentiometer Setting	R	y_o	y_m	θ_1
0 Ω	1000 Ω			
2000 Ω	3000 Ω			
4000 Ω	5000 Ω			

Then compare the measured values of θ_1 obtained in Table 7.3 with those in Table 7.7 by completing Table 7.8, where

$$\% \text{ Difference} = \frac{|\theta_1 \text{ (Table 7.3)} - \theta_1 \text{ (Table 7.7)}|}{\theta_1 \text{ (Table 7.3)}} \times 100\% \qquad (7.13)$$

TABLE 7.8

R	θ_1 (Table 7.3)	θ_1 (Table 7.7)	% Difference
1000 Ω			
3000 Ω			
5000 Ω			

Which method provided the higher degree of accuracy for determining θ_1?

Series Sinusoidal Circuits

OBJECT

To investigate the behavior of series sinusoidal ac circuits at a fixed frequency.

EQUIPMENT REQUIRED

Resistors

1 — 1-kΩ

Inductors

1 — 10-mH

Capacitors

1 — 0.01-μF

Instruments

1 — DMM
1 — Oscilloscope
1 — Audio oscillator or function generator

EQUIPMENT ISSUED

TABLE 8.1

Item	Manufacturer and Model No.	Laboratory Serial No.
DMM		
Oscilloscope		
Audio oscillator or function generator		

TABLE 8.2

Resistors	
Nominal Value	**Measured Value**
1 kΩ	

RÉSUMÉ OF THEORY

Kirchhoff's voltage law is applicable to ac circuits, but it is now stated as follows: *The phasor sum of the voltages around a closed loop is equal to zero.*

For example, in a series circuit, the source voltage is the phasor sum of the component (load) voltages. For a series *R-L-C* circuit,

$$E^2 = V_R^2 + (V_L - V_C)^2$$

so that

$$\boxed{E = \sqrt{V_R^2 + (V_L - V_C)^2}} \tag{8.1}$$

where E is the source voltage, V_R the voltage across the total resistance of the circuit, V_L the voltage across the total inductance, and V_C the voltage across the total capacitance.

Since the resistance and the reactance of a series *R-L-C* circuit are in quadrature,

$$Z^2 = R^2 + X_T^2 \tag{8.2}$$

where X_T(total reactance) $= X_L - X_C$.

In a series *R-L-C* sinusoidal circuit, the voltage across a reactive component may be greater than the input voltage.

The average power (in watts) delivered to a sinusoidal circuit is

$$\boxed{P = I^2R = VI \cos \theta} \tag{8.3}$$

where the values of the current and the voltage are the effective values.

The power factor F_p is given by

$$F_p = \frac{R_T}{Z_T} = \cos \theta \qquad\qquad (8.4)$$

where θ is the angle associated with the total impedance Z_T.

For an ideal inductor, the current lags the voltage across it by 90°. For a capacitor, the current leads the voltage across it by 90°. Inductive circuits are therefore called *lagging power-factor circuits*, and capacitive circuits are called *leading power-factor circuits*.

In a purely resistive circuit, the voltage and the current are in phase and the power factor is unity.

PROCEDURE

Part 1   Series *R-L* Circuit

(a) Construct the series *R-L* network of Fig. 8.1. Insert the measured values of R and R_l.

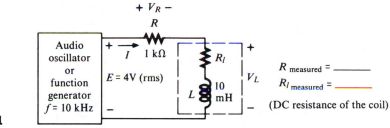

FIG. 8.1

(b) After setting the source voltage E to 4 V (rms) using the DMM, measure the rms values of $\mathbf{V}_R$ and $\mathbf{V}_L$.

$V_R =$ _____ , $V_L =$ _____

(c) Determine the rms value of $\mathbf{I}$ from

$$I_{rms} = \frac{V_{R(rms)}}{R\,(\text{measured})}$$

$I =$ _____

(d) Determine the magnitude of the input impedance from $Z_T = E/I$. Ignore the effects of R_1 since $R \gg R_1$ and $X_L \gg R_1$ at $f = 10$ kHz. In other words, consider the coil to be purely ideal at this frequency $(R_1 \cong 0\ \Omega)$.

$Z_T =$ _____

(e) *Calculate* the magnitude of the total impedance and current I from the nameplate value of the inductance and the measured resistor value. Ignore the effect of R_1. Sketch the impedance diagram and determine the angle associated with the total impedance.

Z_T (theoretical) = _____ , I_{rms} (calculated) = _____

(f) Compare the results of parts 1(d) and 1(e) and explain the source of any difference.

(g) Show that your measurements from part 1(b) satisfy Kirchhoff's voltage law. That is, show that

$$E = \sqrt{V_R^2 + V_L^2}$$

(h) Using $\mathbf{E} = E \angle 0°$ and the angle determined in part 1(e), write the current $\mathbf{I}$ in phasor form.

$I = \underline{\hspace{2cm}}$

(i) Using the fact that $\mathbf{V}_R$ and $\mathbf{I}$ are in phase and $\mathbf{V}_L$ leads $\mathbf{I}$ by 90°, write $\mathbf{V}_R$ and $\mathbf{V}_L$ in phasor form.

$\mathbf{V}_R = \underline{\hspace{2cm}}$

$\mathbf{V}_L = \underline{\hspace{2cm}}$

(j) Sketch the phasor diagram using the results of parts 1(h) and 1(i).

(k) Determine the total power dissipated from $P_R = I^2 R$ and $P = EI \cos \theta$ and compare results.

$P_R =$ _____

$P =$ _____

Part 2   Series *R-C* Circuit

(a) Construct the series *R-C* network of Fig. 8.2. Insert the measured value of *R* in the figure as shown.

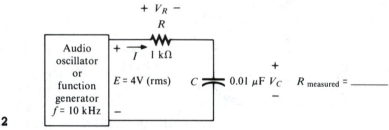

FIG. 8.2

(b) After setting the source voltage *E* to 4 V (rms) using the DMM, measure the rms values of $\mathbf{V}_R$ and $\mathbf{V}_C$.

$V_R =$ _____ , $V_C =$ _____

(c) Determine the rms value of *I* using Eq. (8.5).

$I =$ _____

(d) Determine the magnitude of the input impedance from $Z_T = E/I$.

$Z_T =$ _____

(e) *Calculate* the magnitude of the total impedance and current *I* from the nameplate value of the capacitance and the measured resistor value. Sketch the impedance diagram and determine the angle associated with the total impedance.

Z_T(theoretical) = _____ , $I =$ _____

(**f**) Compare the results of parts 2(d) and 2(e) and explain the sources of any differences.

(**g**) Show that your measurements from part 2(b) satisfy Kirchhoff's voltage law. That is, show that

$$E = \sqrt{V_R^2 + V_C^2}$$

(**h**) Using $\mathbf{E} = E \angle 0°$, and the angle determined in part 2(e), write the current I in phasor form.

$\mathbf{I} =$ _____

(i) Using the fact that $\mathbf{V}_R$ and $\mathbf{I}$ are in phase and $\mathbf{I}$ leads $\mathbf{V}_C$ by 90°, write $\mathbf{V}_R$ and $\mathbf{V}_C$ in phasor form.

$\mathbf{V}_R =$ _____

$\mathbf{V}_C =$ _____

(j) Sketch the phasor diagram using the results of parts 2(h) and 2(i).

Part 3   *R-L-C* Network

(a) Construct the network of Fig. 8.3. Insert the measured resistance values. Ignore the effects of R_l in the following analysis.

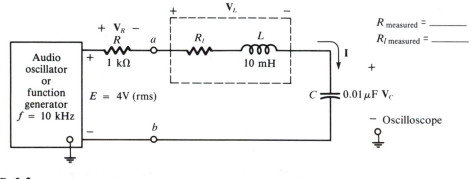

FIG. 8.3

(b) Measure all the component voltages with $E = 4$ V (rms).

$V_R =$ _____ , $V_L =$ _____ ,

$V_C =$ _____

(c) Determine I from $I = V_R/R_{\text{measured}}$.

$I =$ _____

(d) Calculate Z_T from $Z_T = E/I$

$Z_T =$ _____

(e) Calculate the magnitudes of the total impedance and current I from the nameplate values of C and L and the measured resistance level. Sketch the impedance diagram and determine the angle associated with the total impedance.

Z_T (theoretical) $=$ _____ , $I =$ _____

(f) Compare the results of parts 3(d) and 3(e) and explain the source of any differences.

(g) Show that your measurements from part 3(b) satisfy Kirchhoff's voltage law. That is, show that

$$E = \sqrt{V_R^2 + (V_L - V_C)^2}$$

(h) Using $\mathbf{E} = E \angle 0°$, and the angle determined in part 3(e), write the current **I** in phasor form.

$\mathbf{I} = $ _____

(i) Using the fact that $\mathbf{V}_R$ and **I** are in phase, $\mathbf{V}_L$ leads **I** by 90°, and **I** leads $\mathbf{V}_C$ by 90°, write the phasor forms of $\mathbf{V}_R$, $\mathbf{V}_L$, and $\mathbf{V}_C$.

$\mathbf{V}_R = $ _____

$\mathbf{V}_L = $ _____

$\mathbf{V}_C = $ _____

(j) Sketch the phasor diagram using the results of parts 3(h) and 3(i).

Part 4   Phase Angle Measurement

(a) The following procedure will determine the phase angle between the voltage V_C and the source voltage E. Using a dual channel scope, hook up one channel across the source voltage in Fig. 8.3. Adjust the output of the generator to provide a nice 8 V (p-p) sine wave. Then hook up channel 2 across the capacitor with the ground side of the channel connected to the same leg as the ground of the generator.

From the resulting dual channel display, calculate the phase angle between E and V_C using the technique described in Experiment 7.

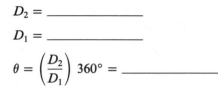

$D_2 = $ _____

$D_1 = $ _____

$\theta = \left(\dfrac{D_2}{D_1}\right) 360° = $ _____

(b) Compare the result of part 4(a) to the angle obtained from the phasor diagram of part 3(j). The angle measured is the angle between **E** and **V**$_C$ on the diagram.

θ [part 3(j)] = _____ , θ [part 4(a)] = _____

(c) Using the result of part 4(a), determine the angle between $\mathbf{V}_R$ and **E** (and between **I** and **E**).

$\theta = $ _____

(d) How does the result of part 4(c) compare with the value obtained from the phasor diagram of part 3(j)?

Parallel Sinusoidal Circuits

OBJECT

To investigate the behavior of parallel sinusoidal ac circuits.

EQUIPMENT REQUIRED

Resistors

1 — 1-kΩ
2 — 10-Ω

Inductors

1 — 10-mH

Capacitors

1 — 0.01-μF

Instruments

1 — DMM
1 — Oscilloscope
1 — Audio oscillator or function generator

EQUIPMENT ISSUED

TABLE 9.1

Item	Manufacturer and Model No.	Laboratory Serial No.
DMM		
Oscilloscope		
Audio oscillator or function generator		

TABLE 9.2

Resistors	
Nominal Value	Measured Value
1 kΩ	
10 Ω	
10 Ω	

RÉSUMÉ OF THEORY

Kirchhoff's current law as applied to ac circuits states that the phasor sum of the currents entering and leaving a node must equal zero. For example, in a parallel circuit, the source current is the phasor sum of the component (load) currents. For a parallel R-L-C circuit, the magnitude of the source current I is given by $I = \sqrt{I_R^2 + I_{X_T}^2}$, where $I_{X_T} = I_{X_L} - I_{X_C}$ (I_{X_L} is the current in the inductive branch, I_{X_C} is the current in the capacitive branch, and I_R is the current in the resistive branch).

In a parallel R-L-C circuit, it is possible for the magnitude of the source current I to be less than a branch current.

Impedances in parallel combine according to the following equation:

$$\frac{1}{\mathbf{Z}_T} = \frac{1}{\mathbf{Z}_1} + \frac{1}{\mathbf{Z}_2} + \frac{1}{\mathbf{Z}_3} \qquad (9.1)$$

In parallel R-L-C circuits, it is possible for the total impedance to be larger than a branch impedance.

To obtain the waveform of a current on an oscilloscope, it is necessary to view the voltage across a resistor. Since the voltage and current of a resistor are related by $E = IR$ (a linear relationship), the current and voltage waveforms are always of the same appearance and in phase.

PROCEDURE

Part 1   *R-L* Parallel Network

      (a) Construct the network of Fig. 9.1. Insert the measured value of each resistor. The magnitudes of R and X_L at the applied frequency permit ignoring (on an approximate basis) the effects of the sensing resistor R_S and the resistance of the inductance R_l when we analyze the system. In other words, assume that you have an ideal parallel *R-L* system. Set the source voltage to 8 V (*p-p*) using the oscilloscope with the scope grounded at point *c*.

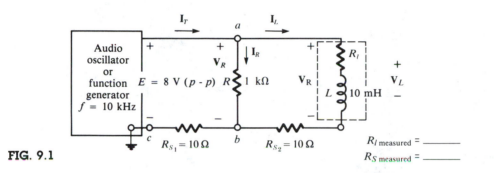

FIG. 9.1

      (b) At the applied frequency, determine the reactance of the coil using the nameplate inductance level of 10 mH.

$$X_L = \underline{\hspace{2in}}$$

      Calculate the peak-to-peak values of I_R and I_L using Ohm's Law. Ignore the effects of R_S.

$$I_{R(p-p)} = \underline{\hspace{1.5in}}$$

$$I_{L(p-p)} = \underline{\hspace{1.5in}}$$

      (c) Calculate the peak-to-peak value of I_T using the results of part 1(c).

$I_{T(p\text{-}p)} =$ _____

(d) Measure the voltages $\mathbf{V}_R$ and $\mathbf{V}_L$ using the oscilloscope. Place the ground of the scope at point b.

$\mathbf{V}_{R(p\text{-}p)} =$ _____

$\mathbf{V}_{L(p\text{-}p)} =$ _____

How do these values compare with the magnitude of the source voltage $\mathbf{E}$?

Is the resistance R_S small enough so that it doesn't affect the relationship between $\mathbf{E}$ and $\mathbf{V}_R$ and $\mathbf{V}_L$?

(e) Place the scope across the sensing resistor R_S with the ground of the scope at point c and the high end of the channel at point b. Measure the peak-to-peak value of the voltage V_{R_S} using the oscilloscope.

$V_{R_S(p\text{-}p)} =$ _____

Calculate the peak-to-peak value of I_T using the above measurement.

$I_{T(p\text{-}p)} =$ _____

How does the result just obtained compare with the calculated value of part 1(c)?

(f) Using $\mathbf{E} = E \angle 0°$, sketch the phasor diagram of $\mathbf{E}$, $\mathbf{I}_R$, $\mathbf{I}_L$, and $\mathbf{I}_T$. Determine the phase angle between $\mathbf{E}$ and $\mathbf{I}_T$.

$\theta_T = $ _____

(g) Calculate the total impedance of the network at $f = 10$ kHz.

$Z_T = $ _____

(h) Calculate the peak-to-peak value of I_T using the results of part 1(g) and $E = 8$ V (p-p). How does the magnitude compare with the results of part 1(e)?

$I_{T(p\text{-}p)} = $ _____

Part 2   R-C Parallel Network

(a) Construct the network of Fig. 9.2. Insert the measured values of the resistors. As in part 1, ignore the magnitude of R_S compared with the other impedances of the network when making your calculations. Use the oscilloscope to set E to 8 V (p-p).

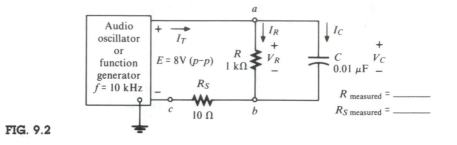

FIG. 9.2

(b) Ignoring the effects of R_S, calculate the total impedance of the network at $f = 10$ kHz using the nameplate value of the capacitance (0.01 μF).

$Z_T = $

(c) Calculate the peak-to-peak value of $\mathbf{I}_T$ using the results of part 2(b).

$\mathbf{I}_{T(p\text{-}p)} = $ _____

(d) Measure the peak-to-peak voltage across the sensing resistor R_S using the oscilloscope. Place the ground of the scope at point c and the higher end of the channel at point b.

$V_{R_{S(p\text{-}p)}} = $ _____

Using the above measurement, calculate the peak-to-peak value of I_T.

$I_{T(p\text{-}p)} = $

(e) Calculate the peak-to-peak values of $\mathbf{I}_R$ and $\mathbf{I}_C$ using Ohm's Law.

$I_{R(p\text{-}p)} = $ _____

$I_{C(p\text{-}p)} = $ _____

Determine the peak-to-peak value of I_T from the above calculations.

$I_{T(p\text{-}p)} = $ _____

How does the magnitude of $I_{T(p\text{-}p)}$ obtained in part 1(d) compare with this result?

(f) Using $E = E \angle 0°$, draw the phasor diagram of $\mathbf{E}$, $\mathbf{I}_R$, $\mathbf{I}_C$, and $\mathbf{I}_T$.

What is the phase angle between $\mathbf{E}$ and $\mathbf{I}_T$?

θ_T (calculated) = _____

(g) The oscilloscope will now be used to determine the phase angle between $\mathbf{E}$ and $\mathbf{I}_T$. The voltage $\mathbf{E}$ will be hooked up to channel 1 as indicated in Fig. 9.3. Even though channel 2 will be used to view the voltage $\mathbf{V}_{R_S}$ since the voltage across a resistor and the current through the resistor are in phase, the phase angle between **E** and V_{R_S} will be the same as between **E** and $\mathbf{I}_T$. Hook up channel 2 as indicated in Fig. 9.3. The corresponding letters can be found in Fig. 9.2. Using the procedure described in an earlier experiment for dual channel scopes, determine the phase angle between **E** and $\mathbf{I}_T$.

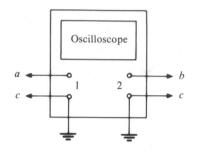

FIG. 9.3

θ_T (measured) = _____

(h) How do the results of parts 2(d) and 2(e) compare?

Part 3 *R-L-C* Parallel Network

(a) Construct the network of Fig. 9.4. Insert the measured resistor values. As in the earlier parts, ignore the sensing resistors and R_l compared with R, X_L, and X_C in your calculations.

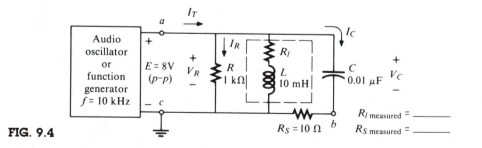

FIG. 9.4

$R_{l \text{ measured}}$ = _____

$R_{S \text{ measured}}$ = _____

(b) Ignoring the effects of R_S and R_l, calculate the total impedance of the network.

$$Z_T = \underline{\hspace{2cm}}$$

(c) Using the results of part 3(b), calculate the magnitude of the peak-to-peak value of $\mathbf{I}_T$.

$$\mathbf{I}_{T(p\text{-}p)} = \underline{\hspace{2cm}}$$

(d) Calculate the peak-to-peak values of I_R, I_L, and I_C.

$$I_{R(p\text{-}p)} = \underline{\hspace{2cm}}$$

$$I_{L(p\text{-}p)} = \underline{\hspace{2cm}}$$

$$I_{C(p\text{-}p)} = \underline{\hspace{2cm}}$$

(e) Using the results of part 3(d), calculate the peak-to-peak value of $\mathbf{I}_T$.

$$\mathbf{I}_{T(p\text{-}p)} = \underline{\hspace{2cm}}$$

How does this value compare with the result of part 3(c)?

(f) Using the above calculations and $\mathbf{E} = \mathbf{E} \angle 0°$, sketch the phasor diagram of $\mathbf{E}, \mathbf{I}_T, \mathbf{I}_R, \mathbf{I}_L,$ and $\mathbf{I}_C$.

What is the phase relationship between $\mathbf{E}$ and $\mathbf{I}_C$?

$\theta = $ _____

What is the phase relationship between $\mathbf{E}$ and $\mathbf{I}_T$?

$\theta_T = $ _____

(g) We will now determine the phase relationship between $\mathbf{E}$ and $\mathbf{I}_C$ using the oscilloscope. Hook up channels 1 and 2 as shown in Fig. 9.5, referencing Fig. 9.4.

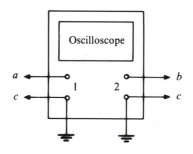

FIG. 9.5

Calculate the phase angle between $\mathbf{E}$ and $\mathbf{I}_C$. How does it compare with the angle obtained in part 3(f)?

$\theta = $ _____

(h) Move the sensing resistor between the resistor R and point c at the bottom of the network and measure the phase angle between $\mathbf{E}$ and $\mathbf{I}_T$.

$\theta_T =$ _____

How does the angle obtained compare with the predicted result of part 3(f)?

(i) With the sensing resistor in the new position, measure V_{R_S} and calculate the peak-to-peak value of I_T.

$I_{T(p-p)} =$ _____

How does this result compare with the calculated levels of parts 3(c) and 3(e)?

Series-Parallel Sinusoidal Circuits

OBJECT

To investigate the behavior of series-parallel sinusoidal networks.

EQUIPMENT REQUIRED

Resistors

1 — 470-Ω, 1-kΩ

Inductors

1 — 10-mH

Capacitors

1 — 0.02-μF

Instruments

1 — DMM

1 — Oscilloscope

1 — Audio oscillator or function generator

EQUIPMENT ISSUED

TABLE 10.1

Item	Manufacturer and Model No.	Laboratory Serial No.
DMM		
Oscilloscope		
Audio oscillator or function generator		

TABLE 10.2

Resistors	
Nominal Value	Measured Value
470 Ω	
1 kΩ	

RÉSUMÉ OF THEORY

In the previous experiments, we showed that Kirchhoff's voltage and current laws hold for ac series and parallel circuits. In fact, all of the previously used rules, laws, and methods of analysis apply equally well for both dc and ac networks. The major difference for ac circuits is that we now use reactance, resistance, and impedance instead of solely resistance.

Consider the common series-parallel circuit of Fig. 10.1.

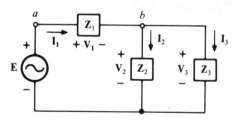

FIG. 10.1

The various impedances shown could be made up of many elements in a variety of configurations. No matter how varied or numerous the elements might be, Z_1, Z_2, and Z_3 represent the total impedance for that branch. For example, Z_1 might be as shown in Fig. 10.2.

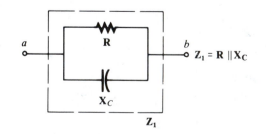

FIG. 10.2

The currents and voltages in Fig. 10.1 can be found by applying any of the methods outlined for dc networks; in particular,

$$\mathbf{Z}_T = \mathbf{Z}_1 + \mathbf{Z}_2 \| \mathbf{Z}_3 \qquad \mathbf{I}_T = \mathbf{I}_1 = \frac{\mathbf{E}}{\mathbf{Z}_T}$$

and

$$\mathbf{I}_2 = \frac{\mathbf{Z}_3 \mathbf{I}_T}{\mathbf{Z}_3 + \mathbf{Z}_2} \qquad \text{(current divider rule)}$$

$\mathbf{I}_3$ can be determined using Kirchhoff's current law:

$$\mathbf{I}_3 = \mathbf{I}_T - \mathbf{I}_2$$

Ohm's Law provides the voltages:

$$\mathbf{V}_1 = \mathbf{I}_1 \mathbf{Z}_1 \qquad \mathbf{V}_2 = \mathbf{I}_2 \mathbf{Z}_2 \qquad \mathbf{V}_3 = \mathbf{I}_3 \mathbf{Z}_3$$

Keep in mind that the voltages and currents are phasor quantities that have both magnitude and an associated angle.

The power factor $\cos \theta = R_T/Z_T$, where R_T is the equivalent total resistance of the circuit at the input terminals.

PROCEDURE

Part 1 *R-L Series-Parallel Network*

(a) Construct the network of Fig. 10.3. Insert the measured value for each resistor. At the applied frequency, the resistor R_l can be ignored in comparison with the other elements of the network. Set the source to 8 V (*p-p*) with the oscilloscope.

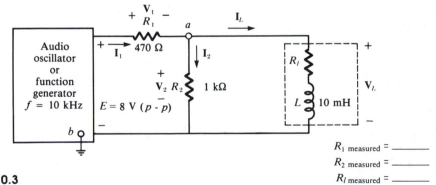

FIG. 10.3

R_1 measured = _____
R_2 measured = _____
R_l measured = _____

(b) Calculate the total impedance of the network at $f = 10$ kHz. Use measured resistor values.

$$Z_T = \text{_____}$$

(c) Using the results of part 1(b), calculate the peak-to-peak value of the source current $I_T = I_1$.

$$I_{T(p\text{-}p)} = \text{_____}$$

(d) Measure the voltage V_1 with the DMM. Convert the reading to a peak-to-peak value and calculate the peak-to-peak value of $\mathbf{I}_1 = \mathbf{I}_T$.

$$V_{1(\text{rms})} \text{ (measured)} = \text{_____}$$

$$V_{1(p\text{-}p)} = \text{_____}$$

$$I_{1(p\text{-}p)} = I_{T(p\text{-}p)} = \text{_____}$$

How does the magnitude of $I_{T(p\text{-}p)}$ compare with the results of part 1(c)?

(e) Measure the peak-to-peak values of $\mathbf{V}_2$ and $\mathbf{V}_L$ with the oscilloscope by connecting the ground terminal of the scope channel to point b and the high end of the channel to point a.

$V_{2(p\text{-}p)} = \underline{\hspace{3cm}}$

$V_{L(p\text{-}p)} = \underline{\hspace{3cm}}$

(f) Using the results of part 1(e), calculate the peak-to-peak values of $\mathbf{I}_2$ and $\mathbf{I}_L$.

$I_{2(p\text{-}p)} = \underline{\hspace{3cm}}$

$I_{L(p\text{-}p)} = \underline{\hspace{3cm}}$

(g) Using the results of parts 1(d) and 1(f), determine whether the following relationship is satisfied:

$$I_1 = \sqrt{I_2^2 + I_L^2}$$

(h) Devise a method to measure the phase angle between $\mathbf{E}$ and $\mathbf{I}_T = \mathbf{I}_1$. Check your method with the instructor and then proceed to determine the angle.

$\theta_T = \underline{\hspace{3cm}}$

(i) Devise a method to measure the phase angle between **E** and V_2 ($= V_L$). Check your method with the instructor and then proceed to determine the angle.

$\theta =$ _____

(j) Using the results of part 1(h), calculate the phase angle between **E** and V_1.

$\theta =$ _____

(k) Draw the phasor diagram of **E**, V_1, V_2, and I_T using the above results and peak-to-peak values.

Part 2   *R-C* Series-Parallel Network

(a) Construct the network of Fig. 10.4. Insert the measured resistor values. Set the source to 8 V (*p-p*) with the oscilloscope.

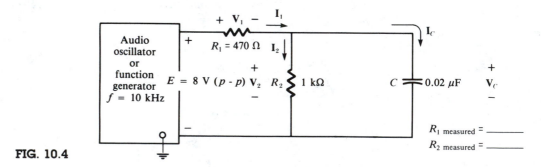

FIG. 10.4

(b) Calculate the magnitude of the total impedance of the network at $f = $ 10 kHz.

$Z_T = $ _____

(c) Measure the rms value of V_1 with the DMM.

$V_{1(rms)} = $ _____

Calculate the peak-to-peak value of V_1.

$V_{1(p-p)} = $ _____

Calculate the magnitude of Z_T from

$$Z_T = \frac{E_{p-p}}{I_{T(p-p)}}$$

using the fact that $I_{T(p-p)} = I_{1(p-p)} = [V_{1(p-p)}/R_1]$

$I_{T(p-p)} = $ _____

$Z_T = $ _____

How do the results for Z_T from parts 2(b) and 2(c) compare?

(d) Measure the peak-to-peak values of $\mathbf{V_2}$ and $\mathbf{V_C}$ with the oscilloscope.

$V_{2(p-p)} = $ _____

$V_{C(p-p)} = $ _____

Calculate the peak-to-peak values of I_2 and I_C.

$$I_{2(p\text{-}p)} = \underline{\hspace{2cm}}$$

$$I_{C(p\text{-}p)} = \underline{\hspace{2cm}}$$

(e) Using the results of parts 2(c) and 2(d), is the following relationship satisfied?

$$I_1 = \sqrt{I_2^2 + I_C^2}$$

(f) Devise a method to measure the phase angle between $\mathbf{E}$ and $\mathbf{I}_T$. Check the method with your instructor and then proceed to determine the angle.

$$\theta_T = \underline{\hspace{2cm}}$$

(g) Devise a method to measure the phase angle between $\mathbf{E}$ and $\mathbf{I}_2$. Check the method with your instructor and then proceed to determine the angle.

$$\theta = \underline{\hspace{2cm}}$$

(h) Devise a method to determine the phase angle between $\mathbf{I}_2$ and $\mathbf{I}_C$. Check the method with your instructor and then proceed to determine the angle.

$\theta = $ _____

(i) If $\mathbf{E} = E \angle 0°$, write $\mathbf{V}_1$ and $\mathbf{V}_2$ in phasor form using the results obtained above. Use peak-to-peak values for the magnitudes.

$\mathbf{V}_1 = $ _____

$\mathbf{V}_2 = $ _____

(j) Is the following Kirchhoff's voltage law relationship satisfied?

$$E = \sqrt{V_1^2 + V_2^2}$$

Thevenin's Theorem and Maximum Power Transfer

OBJECT

To study Thevenin's theorem and the theorem of maximum power transfer as they apply to ac circuits and sources.

EQUIPMENT REQUIRED

Resistors

1 — 470-Ω, 1-kΩ, 2.2-kΩ

1 — 0–1-kΩ potentiometer

Capacitors

1 — 0.0047-μF, 0.01-μF, 0.1-μF

2 — 0.02-μF

Inductors

1 — 10-mH

Instruments

1 — DMM

1 — Oscilloscope

1 — Audio oscillator or function generator

EQUIPMENT ISSUED

TABLE 11.1

Item	Manufacturer and Model No.	Laboratory Serial No.
DMM		
Oscilloscope		
Audio oscillator or function generator		

TABLE 11.2

Resistors	
Nominal Value	**Measured Value**
470 Ω	
1 kΩ	
2.2 kΩ	

RÉSUMÉ OF THEORY

Thevenin's theorem states that any two-terminal linear ac network can be replaced by an equivalent circuit consisting of a voltage source in series with an impedance. To apply this theorem, follow these simple steps:

1. Remove the portion of the network across which the Thevenin equivalent circuit is found.
2. Short out all voltage sources and open all current sources. (This is never done practically, but only theoretically.)
3. Calculate Z_{Th} across the two terminals in question.
4. Replace all sources.
5. Calculate E_{Th}, which is the voltage across the terminals in question.
6. Draw the Thevenin equivalent circuit and replace the portion of the circuit originally removed.
7. Solve for the voltage or current originally desired.

EXAMPLE   Using Thevenin's theorem, find the current through R in Fig. 11.1.

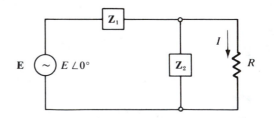

FIG. 11.1

Step 1: Remove R. See Fig. 11.2(a).
Step 2: Replace **E** by a short-circuit equivalent. See Fig. 11.2(a).
Step 3: Solve for $\mathbf{Z}_{Th}$. In this case, $\mathbf{Z}_{Th} = \mathbf{Z}_1 || \mathbf{Z}_2$. See Fig. 11.2(a).
Step 4: Replace **E**. See Fig. 11.2(b).
Step 5: Calculate $\mathbf{E}_{Th}$ from

$$\mathbf{E}_{Th} = \frac{\mathbf{Z}_2}{\mathbf{Z}_1 + \mathbf{Z}_2} \mathbf{E}$$

See Fig. 11.2(b).
Step 6: Draw the equivalent circuit and replace R. See Fig. 11.2(c).
Step 7: Calculate **I** from

$$\mathbf{I} = \frac{\mathbf{E}_{Th}}{\mathbf{Z}_{Th} + \mathbf{R}}$$

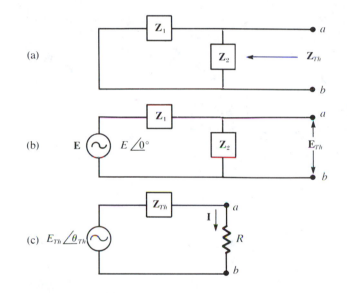

FIG. 11.2

MAXIMUM POWER TRANSFER THEOREM

The maximum power transfer theorem states that for circuits with ac sources, maximum power will be transferred to a load when the load impedance is the conjugate of the Thevenin impedance across its terminals. See Fig. 11.3.

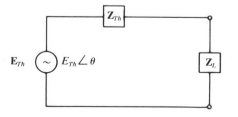

FIG. 11.3

For maximum power transfer, if

$$\mathbf{Z}_{Th} = |Z_{Th}| \angle \theta$$

then

$$\mathbf{Z}_L = |Z_{Th}| \angle -\theta$$

PROCEDURE

Part 1

(a) Measure the dc resistance of the coil and include it in Fig. 11.4. Indicate the measured values of R, R_1, and R_2 from Table 11.2.

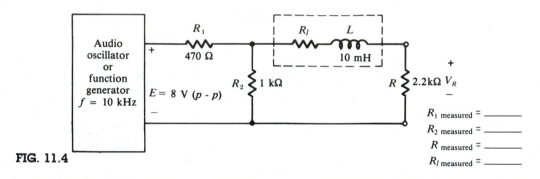

FIG. 11.4

(b) With E set at 8 V (p-p), measure $V_{R(p\text{-}p)}$ and calculate $V_{R(\text{rms})}$.

$V_{R(p\text{-}p)} =$ _____ , $V_{R(\text{rms})} =$ _____

(c) Remove the resistor R and measure the open-circuit voltage across the resulting terminals. This is the magnitude of $\mathbf{E}_{Th}$.

$|\mathbf{E}_{Th}|_{p\text{-}p} =$ _____ , $|\mathbf{E}_{Th}|_{\text{rms}} =$ _____

(d) Using the measured resistance levels, calculate the magnitude of $\mathbf{E}_{Th}$ and its associated angle. Compare with the magnitude of part 1(c). Assume that $\mathbf{E} = E \angle 0° = 0.707(8/2) \angle 0° = 2.828$ V $\angle 0°$.

$\mathbf{E}_{Th} =$ _____

(e) Calculate $\mathbf{Z}_{Th}$ (magnitude and angle) using the measured resistor values and the nameplate inductor level of 10 mH. The applied frequency is 10 kHz.

$\mathbf{Z}_{Th} =$ _____

Determine the resistive and reactive components of $\mathbf{Z}_{Th}$.

$R =$ _____ , $X_L =$ _____

(f) Insert the calculated values obtained for $\mathbf{E}_{Th}$ and $\mathbf{Z}_{Th}$ in parts 1(c) and 1(e) in Fig. 11.5. Using the 1-kΩ potentiometer and the amplitude control on the audio oscillator or function generator, set the values of the Thevenin circuit. Calculate the inductance level from the results of part 1(e) and the fact that $L = X_L/\omega$. Then measure the voltage $V_{R(p\text{-}p)}$ and compare with the measurement of part 1(b).

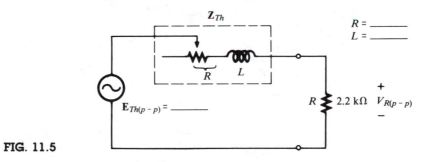

FIG. 11.5

$V_{R(p\text{-}p)} =$ _____ , $V_{R(rms)} =$ _____

(g) Is the Thevenin theorem verified?

Part 2 Maximum Power Transfer

(a) Construct the circuit of Fig. 11.6. Include the measured resistor values.

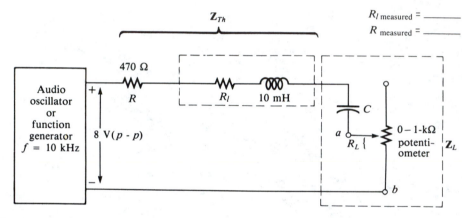

FIG. 11.6

(b) Set the 0–1-kΩ potentiometer to $R_T = R_{measured} + R_l$ as required for maximum power transfer. Indicate this value in each row of Table 11.3.

$R_T = R\,(\text{measured}) + R_l =$ _____ + _____ = _____

TABLE 11.3

R_T	X_L	$|Z_{Th}|$	$C\,(\mu F)$	X_C	$|Z_L|$	$V_{ab(rms)}$	$P_L = V_{ab}^2/R_T$
			0.0047				
			0.01				
			0.02				
			0.0247				
			0.05				
			0.1				

(c) For $f = 10$ kHz, calculate X_L (ignore any effects of R_l) and insert (the same value) in each row of Table 11.3.

(d) Using the results of parts 2(b) and 2(c), calculate the magnitude of the Thevenin impedance and insert (the same value) in each row of Table 11.3.

(e) For each value of C, calculate X_C and insert in Table 11.3.

(f) Calculate the magnitude of $\mathbf{Z}_L$ and insert in each row of Table 11.3

(g) Energize the network and, for each capacitance value, measure the rms value of V_{ab} with the DMM and insert in Table 11.3. Calculate the real power to the load using $P_L = V_{ab}^2/R_T$ and insert in Table 11.3. (Note that $0.0247 \ \mu F = 0.02 \ \mu F + 0.0047 \ \mu F$.)

(h) Plot P_L versus X_C on Graph 11.1.

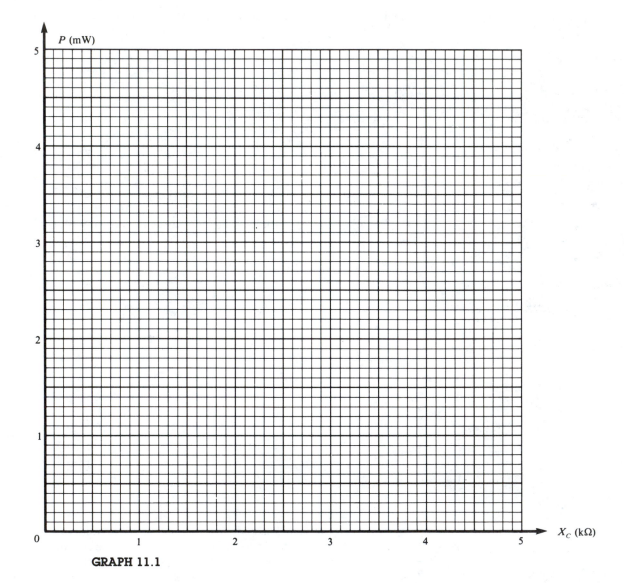

GRAPH 11.1

(i) Is maximum power transferred when $R_L = R_T$ and $X_C = X_L$ as required by the theorem?

(j) Assuming that L is exactly 10 mH, what value of capacitance will ensure that $X_L = X_C$ at $f = 10$ kHz? How close do we come to this value in Table 11.3?

Series Resonant Circuits

OBJECT

To investigate the characteristics of a series resonant circuit.

EQUIPMENT REQUIRED

Resistors

1 — 47-Ω, 220-Ω

Inductors

1 — 1-mH, 10-mH

Capacitors

1 — 0.1-μF, 1-μF

Instruments

1 — DMM
1 — Audio oscillator or function generator

EQUIPMENT ISSUED

TABLE 12.1

Item	Manufacturer and Model No.	Laboratory Serial No.
DMM		
Audio oscillator or function generator		

TABLE 12.2

Resistors	
Nominal Value	**Measured Value**
47 Ω	
220 Ω	

RÉSUMÉ OF THEORY

In a series R-L-C circuit, there exists one frequency at which $X_L = X_C$ or $\omega L = 1/\omega C$. At this frequency the circuit is in resonance, and the input voltage and current are in phase. At resonance, the circuit is resistive in nature and has a minimum value of impedance or a maximum value of current.

The resonant radian frequency $\omega_s = 1/\sqrt{LC}$ and frequency $f_s = 1/2\pi\sqrt{LC}$. The Q of the circuit is defined as $\omega_s L/R$ and affects the selectivity of the circuit. High-Q circuits are very selective. At resonance, $V_C = QE_{input}$. The half-power frequencies f_1 and f_2 are defined as those frequencies at which the power dissipated is one-half the power dissipated at resonance. In addition, the current is 0.707 (or $1/\sqrt{2}$) times the current at resonance.

The bandwidth BW $= f_2 - f_1$. The smaller the bandwidth, the more selective the circuit. In Fig. 12.1, note that increasing R results in a less selective circuit. Figure 12.2

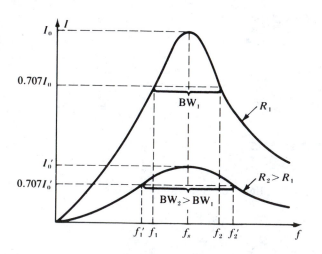

FIG. 12.1

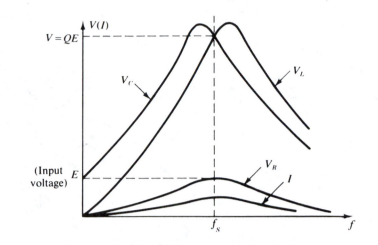

FIG. 12.2

shows the voltages across the three elements versus the frequency. The voltage across the resistor, V_R, has exactly the same shape as the current since it differs by the constant R. V_R is a maximum at resonance. V_C and V_L are equal at resonance (f_s) since $X_L = X_C$, but note that they are not maximum at the resonant frequency. At frequencies below f_s, $V_C > V_L$; at frequencies above f_s, $V_L > V_C$, as indicated in Fig. 12.2.

PROCEDURE

Part 1 Low-Q Circuit

(a) Construct the circuit of Fig. 12.3. Insert the measured resistance values.

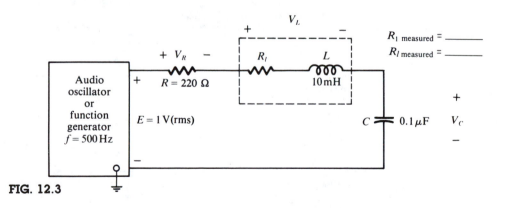

FIG. 12.3

(b) Using the nameplate values and measured resistance values, compute the radian frequency ω_s, the frequency f_s, and the Q_s of the circuit at resonance. Include the effects of R_l in your calculations.

$\omega_s =$ _____ , $f_s =$ _____ , $Q_s =$ _____

(c) Energize the circuit and set the oscillator to the frequencies indicated in Table 12.3. At each frequency, reset the input to 1 V (rms) with the DMM and measure the rms values of the voltages V_C, V_R, and V_L with the DMM. Take a few extra readings near the resonant frequency to improve your data.

TABLE 12.3

Frequency	$V_{R\,(rms)}$	$V_{L\,(rms)}$	$V_{C\,(rms)}$	$I_{rms} = \dfrac{V_{R\,(rms)}}{R}$
500 Hz				
1,000 Hz				
2,000 Hz				
3,000 Hz				
4,000 Hz				
5,000 Hz				
6,000 Hz				
7,000 Hz				
8,000 Hz				
9,000 Hz				
10,000 Hz				
f_s(calculated), Hz				
Near $\{$ f_s $\{$				

Part 2 Higher-Q Circuit

(a) Repeat parts 1(a) through 1(c), replacing the 220-Ω resistor with a resistor of 47 Ω. Record the results in Table 12.4.

$\omega_s =$ _____ , $f_s =$ _____ , $Q_s =$ _____

TABLE 12.4

Frequency	$V_{R(rms)}$	$V_{L(rms)}$	$V_{C(rms)}$	$I_{rms} = \dfrac{V_{R(rms)}}{R}$
500 Hz				
1,000 Hz				
2,000 Hz				
3,000 Hz				
4,000 Hz				
5,000 Hz				
6,000 Hz				
7,000 Hz				
8,000 Hz				
9,000 Hz				
10,000 Hz				
f_s(calculated), Hz				
Near f_s {				

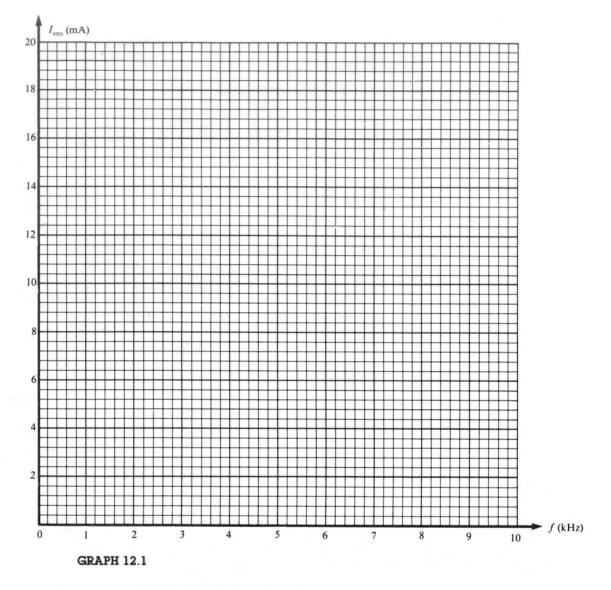

GRAPH 12.1

(b) On Graph 12.1, draw the curves of current I_{rms} (in milliamperes) versus frequency for parts 1 and 2. Label each curve.

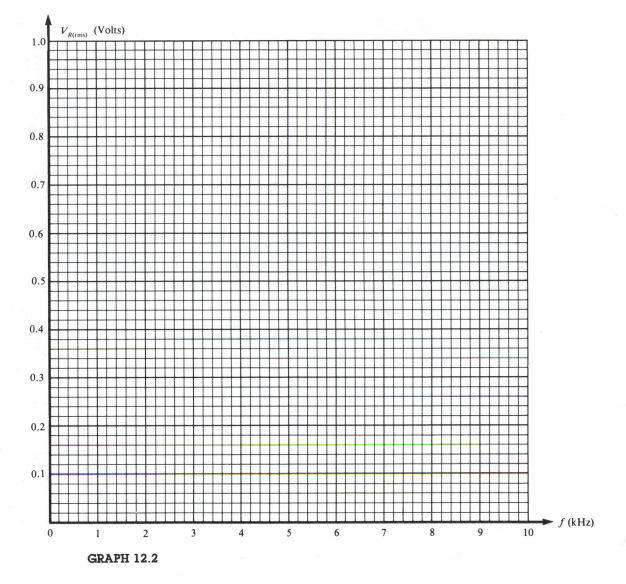

GRAPH 12.2

(c) On Graph 12.2, plot $V_{R(rms)}$ versus frequency for parts 1 and 2. Label each curve.

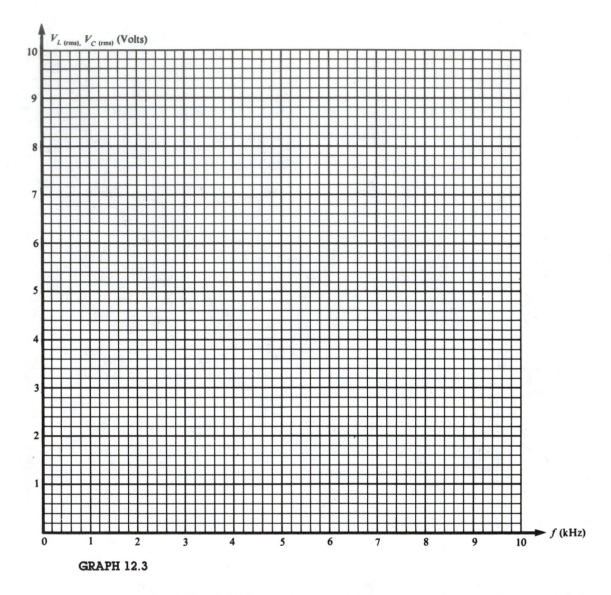

GRAPH 12.3

   (d) On Graph 12.3, plot $V_{L(rms)}$ and $V_{C(rms)}$ versus frequency for parts 1 and 2. Label each curve.

   (e) How do the curves of parts 2(b) through 2(d) compare with those in the Résumé of Theory?

(f) Indicate the resonant and half-power frequencies and the bandwidth on each curve. What is the bandwidth in hertz for each curve using the results of part 2(b)?

BW $(R = 220\ \Omega) = $ _____ , BW $(R = 47\ \Omega) = $ _____

Which curve is more selective? Why?

(g) Calculate the Q_s for each part from the data *at the resonant frequency*. That is, use the peak-to-peak values of V_C and E and the following equation:

$$Q_s = \frac{V_C}{E}$$

$Q_s(R = 220\ \Omega) = $ _____ , $Q_s(R = 47\ \Omega) = $ _____

(h) Calculate the Q_s of each circuit from the curves using $Q_s = f_s/\text{BW}$. How do the Q_s's obtained in parts 2(g) and 2(h) compare?

$Q_s(R = 220\ \Omega) = $ _____ , $Q_s(R = 47\ \Omega) = $ _____

Parallel Resonant Circuits

OBJECT

To investigate the characteristics of a parallel resonant circuit.

EQUIPMENT REQUIRED

Resistors

$1-22\text{-}\Omega$, 10-kΩ

Inductors

$1-1$-mH

Capacitors

$1-1$-μF

Instruments

$1-$DMM

$1-$Oscilloscope

$1-$Audio oscillator or function generator

EQUIPMENT ISSUED

TABLE 13.1

Item	Manufacturer and Model No.	Laboratory Serial No.
DMM		
Oscilloscope		
Audio oscillator or function generator		

TABLE 13.2

Resistors	
Nominal Value	Measured Value
22 Ω	
10 kΩ	

RÉSUMÉ OF THEORY

The basic components of a parallel resonant network appear in Fig. 13.1.

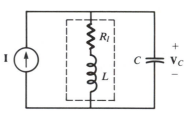

FIG. 13.1

For parallel resonance, the following condition must be satisfied:

$$\frac{1}{X_C} = \frac{X_L}{X_L^2 + R_l^2}$$

(13.1)

At resonance, the impedance of the network is resistive and is determined by

$$Z_{T_p} = \frac{L}{R_l C}$$

(13.2)

The resonant frequency is determined by

$$f_p = \frac{1}{2\pi\sqrt{LC}}\sqrt{1 - \frac{R_l^2 C}{L}}$$

(13.3)

which reduces to

$$\boxed{f_p = \frac{1}{2\pi\sqrt{LC}}}$$

(13.4)

when

$$Q_{coil} = \frac{X_L}{R_l} \geq 10$$

For parallel resonance, the curve of interest is that of V_C versus frequency due to electronic considerations that often place this voltage at the input to the following stage. It has the same shape as the resonance curve for the series configuration with the bandwidth determined by

$$\boxed{BW = f_2 - f_1}$$

(13.5)

where f_2 and f_1 are the cutoff or band frequencies.

PROCEDURE

Part 1   High-Q Parallel Resonant Circuit

(a) Construct the network of Fig. 13.2. Insert the measured resistance values.

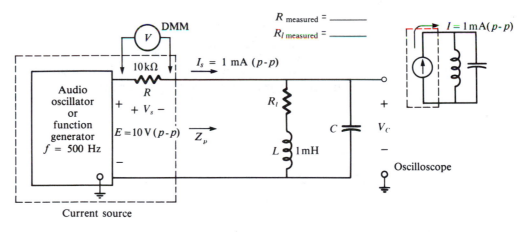

FIG. 13.2

(b) The input voltage E must be maintained at 10 V (p-p). Make sure to reset E after each new frequency is set on the oscillator.

Since the input to a parallel resonant circuit is usually a constant-current device such as a transistor, the input current I_s has been designed to be fairly constant in magnitude for the frequency range of interest. The input impedance (Z_p) of the parallel resonant circuit will always be sufficiently less than the 10-kΩ resistor, to permit the following approximation:

$$I_s = \frac{E}{10 \text{ k}\Omega + Z_p} \cong \frac{E}{10 \text{ k}\Omega} = \frac{10 \text{ V } (p\text{-}p)}{10 \text{ k}\Omega} = 1 \text{ mA } (p\text{-}p)$$

(c) Compute the resonant frequency of the network of Fig. 13.2 using the nameplate data and the measured resistor values.

$f_p =$ _____

(d) Calculate the input impedance at resonance using the following equation:

$$Z_p = R_p = \frac{L}{R_l C}$$

$Z_p =$ _____

(e) Compare the 10-kΩ resistor with the result of part 1(c). Is it reasonable to assume that the input current is fairly constant for the frequency range of interest?

(f) Is this a high- or low-Q circuit? Find Q_p using

$$Q_p = \frac{X_L}{R_l}$$

$Q_p =$ _____

(g) Vary the frequency from 500 Hz to 10 kHz and use the oscilloscope to measure $V_{C(p\text{-}p)}$ at each frequency. Make sure to maintain E at 10 V (p-p) for each frequency. Insert in Table 13.3. At each frequency, measure $V_{s(\text{rms})}$ with the DMM and calculate $V_{s(p\text{-}p)}$ from $V_{s(p\text{-}p)} = 2(1.414 V_{s(\text{rms})}) = 2.828 V_{s(\text{rms})}$. Insert $V_{s(p\text{-}p)}$ in Table 13.3. Calculate $I_{s(p\text{-}p)}$ from $I_{s(p\text{-}p)} = V_{s(p\text{-}p)}/R_s$ at each frequency and complete Table 13.3. Take a few extra readings around f_p.

TABLE 13.3

Frequency	$V_{C(p\text{-}p)}$	$V_{s(p\text{-}p)}$	$I_{s(p\text{-}p)}$
500 Hz			
1,000 Hz			
2,000 Hz			
3,000 Hz			
4,000 Hz			
5,000 Hz			
6,000 Hz			
7,000 Hz			
8,000 Hz			
9,000 Hz			
10,000 Hz			
f_p(calculated), Hz			
Near f_p {			
{			

(h) Plot $V_{C(p-p)}$ versus frequency on Graph 13.1. The choice of vertical scaling is part of the exercise.

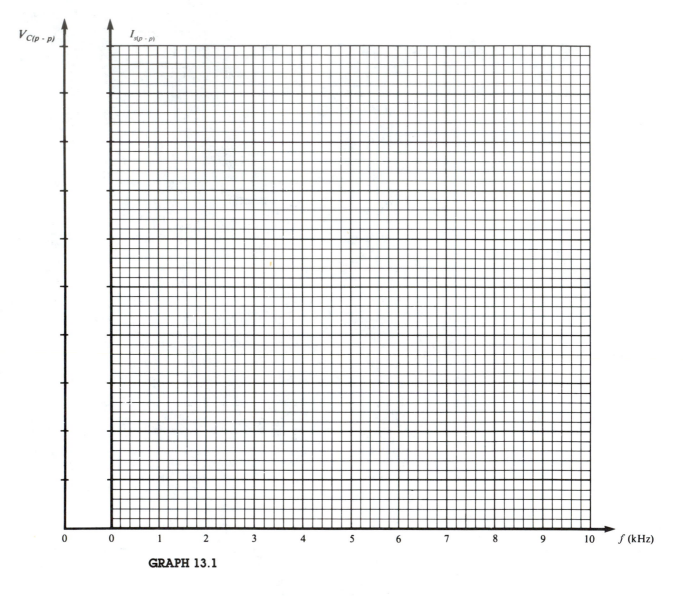

GRAPH 13.1

(i) Plot the curve of $I_{s(p-p)}$ versus f on Graph 13.1. The choice of vertical scaling is left as an exercise.

(j) Is $I_{s(p-p)}$ fairly constant for the frequency range of interest?

(k) Using the results of part 1(h), determine the bandwidth and cutoff frequencies.

BW = _____ , f_1 = _____ , f_2 = _____

(l) Compute the bandwidth from BW = f_p/Q_p and compare with the result of part 1(k).

$BW_{computed}$ = _____ , $BW_{measured}$ [part 1(k)] = _____

Part 2   Lower-Q Parallel Resonant Circuit

(a) Insert the 22-Ω resistor in series with the inductor of Fig. 13.2 so that the total resistance of the inductor branch is R_l + 22 Ω (measured value). The effect will be to reduce the Q_p of the network as determined by $Q_p = X_L/R$.

Determine the resonant frequency of the network using Eq. (13.3). The value of R in the equation must now include the 22-Ω resistor.

f_p = _____

(b) Calculate the input impedance at resonance.

$$Z_{T_p} = \underline{\hspace{2cm}}$$

(c) Calculate the Q_p of the network at the resonant frequency. Compare with the result of part 1(f).

$$Q_p = \underline{\hspace{2cm}} , \quad Q_p \text{ [from part 1(f)]} = \underline{\hspace{2cm}}$$

(d) Repeat part 1(g) and insert the data in Table 13.4. Take a few extra readings around the resonant frequency.

TABLE 13.4

Frequency	$V_{C(p\text{-}p)}$	$V_{s(p\text{-}p)}$	$I_{s(p\text{-}p)}$
500 Hz			
1,000 Hz			
2,000 Hz			
3,000 Hz			
4,000 Hz			
5,000 Hz			
6,000 Hz			
7,000 Hz			
8,000 Hz			
9,000 Hz			
10,000 Hz			
f_p(calculated), Hz			
Near f_p			

(e) Plot $V_{C(p\text{-}p)}$ on Graph 13.1 and compare with the results of part 1(h).

(f) Determine the bandwidth and cutoff frequencies from the plot of part 2(e).

BW = _____ , f_2 = _____ , f_1 = _____

The Transformer

OBJECT

To study the characteristics of a transformer.

EQUIPMENT REQUIRED

Resistors

1 — 100-Ω, 220-Ω, 10-kΩ

Capacitors

1 — 15-μF

Transformers

Filament — 120-V primary, 12.6-V secondary

Instruments

1 — DMM

1 — Oscilloscope

1 — Audio oscillator or function generator

EQUIPMENT ISSUED

TABLE 14.1

Item	Manufacturer and Model No.	Laboratory Serial No.
DMM		
Oscilloscope		
Audio oscillator or function generator		

TABLE 14.2

Resistors	
Nominal Value	**Measured Value**
100 Ω	
220 Ω	
10 kΩ	

RÉSUMÉ OF THEORY

A *transformer* is a device that transfers energy from one circuit to another by electromagnetic induction. The energy is always transferred without a change in frequency. The winding connected to the energy source is called the *primary*, while the winding connected to the load is called the *secondary*. A step-up transformer receives electrical energy at one voltage and delivers it at a higher voltage. Conversely, a step-down transformer receives energy at one voltage and delivers it at a lower voltage. Transformers require little care and maintenance because of their simple, rugged, and durable construction. The efficiency of power transformers is also quite high.

The operation of the transformer is based on the principle that electrical energy can be transferred efficiently by mutual induction from one winding to another.

The physical construction of a transformer is shown in Fig. 14.1.

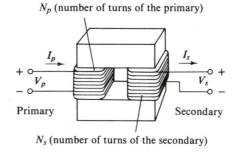

N_p (number of turns of the primary)

I_p I_s

$+$ $+$
V_p V_s
$-$ $-$

Primary   Secondary

FIG. 14.1 N_s (number of turns of the secondary)

For practical applications, the apparent power at the input to the primary circuit ($S = V_p I_p$) is equal to the apparent power rating at the secondary ($S = V_s I_s$). That is,

$$\boxed{V_p I_p = V_s I_s}$$ **(14.1)**

It can also be shown that the ratio of the primary voltage V_p to the secondary voltage V_s is equal to the ratio of the number of turns of the primary to the number of turns of the secondary:

$$\boxed{\frac{V_p}{V_s} = \frac{N_p}{N_s}}$$ **(14.2)**

The currents in the primary and secondary circuits are related in the following manner:

$$\boxed{\frac{I_p}{I_s} = \frac{N_s}{N_p}}$$ **(14.3)**

The impedance of the secondary circuit is electrically reflected to the primary circuit as indicated by the following expression:

$$\boxed{Z_p = a^2 Z_s}$$ **(14.4)**

where $a = N_p/N_s$.

PROCEDURE

Part 1   Voltage, Current, and Turns Ratio of the Transformer

(a) Construct the circuit of Fig. 14.2. Insert the measured resistor values. Set the input voltage to 20 V (p-p) using the oscilloscope.

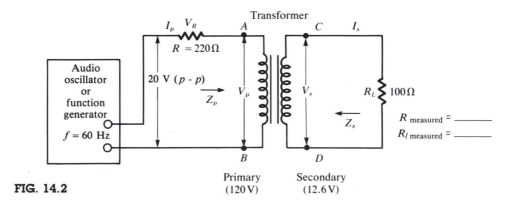

FIG. 14.2

Measure and record the primary and secondary rms voltages and V_R with the DMM.

$$V_{p(\text{rms})} = \text{_____} \; , \quad V_{s(\text{rms})} = \text{_____} \; ,$$

$$V_{R(\text{rms})} = \text{_____}$$

Using the measured values above and Eq. (14.2), calculate the turns ratio N_p/N_s.

$$a = N_p/N_s = \text{_____}$$

(b) Determine the current I_s shown in Fig. 14.2 using Ohm's Law ($I_s = V_s/R_L$). Using Eq. (14.3), calculate I_p.

$$I_{s(\text{rms})} \,(\text{measured}) = \text{_____} \; , \quad I_{p(\text{rms})} \,(\text{calculated}) = \text{_____}$$

(c) Determine the current I_p shown in Fig. 14.2 using Ohm's Law ($I_p = V_R/R_s$).

$$I_{p(\text{rms})} = \text{_____}$$

How does this measured value compare with the calculated value of part 1(b)? Is this a step-up or step-down transformer? Explain.

(d) Set up the network of Fig. 14.3. Note the new position of terminals A through D. In essence, the transformer is hooked up in the reverse manner. Insert the measured resistor values. Set the input voltage to 2 V (p-p) using the oscilloscope.

Now measure the rms values of V_p, V_s, and V_R. Use the technique of the previous parts to determine I_p and I_s.

$$V_{p(\text{rms})} = \text{_____} \; , \quad I_{p(\text{rms})} = \text{_____}$$

$$V_{s(\text{rms})} = \text{_____} \; , \quad I_{s(\text{rms})} = \text{_____}$$

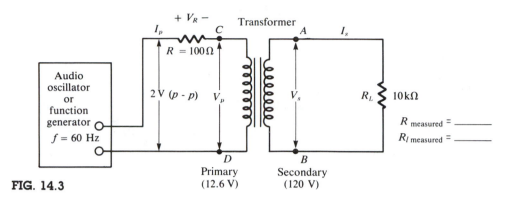

FIG. 14.3

Using Eqs. (14.2) and (14.3), calculate the turns ratio. Is the transformer being used as a step-up or step-down transformer? Explain.

N_p/N_s [Eq. (14.2)] = _____ , N_p/N_s [Eq. (14.3)] = _____

Part 2   Power Dissipation

Using the measured values of I_p, V_p, I_s, and V_s, calculate the apparent power ($P = VI$) of the primary and secondary circuits of the transformer of Fig. 14.2.

$P_{primary}$ = _____ , $P_{secondary}$ = _____

What conclusion can you draw from your calculations?

Part 3   Impedance Transformation

Reconnect the transformer as shown in Fig. 14.2.

   (a) Using Eq. (14.4), calculate the reflected impedance (Z_p) for the network of Fig. 14.2. Do not include R in Z_p.

Z_p(calculated) = _____

(b) Calculate Z_p from the data of parts 1(a) and 1(c). That is, use $Z_p = V_p/I_p$.

$Z_p = $ _____

How does this value of Z_p compare with the calculated value in part 3(a)?

(c) Disconnect the circuit (Fig. 14.2) from the oscillator or generator. Connect the 15-μF capacitor in series with the 100-Ω resistor (see Fig. 14.4). Calculate the magnitude of the reflected impedance Z_p for this circuit.

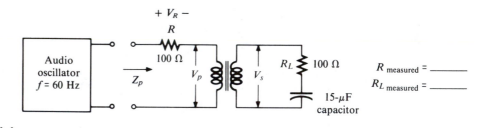

FIG. 14.4

$Z_p = $ _____

(d) Connect the circuit to the oscillator or generator and set the oscillator (or generator) to 2 V (p-p). Measure V_R and V_p with the DMM.

$V_R = $ _____ , $V_p = $ _____

Calculate I_p from $I_p = V_R/R$.

$I_p =$ _____

Calculate Z_p (the magnitude) using the values of I_p and V_p above.

$Z_p =$ _____

How does this value of Z_p compare with the value calculated in part 3(b)?

Power, and Voltage Measurements in Three-Phase Balanced Systems

OBJECT

To investigate the current and voltage relations in three-phase systems and the use of the two-wattmeter method for balanced systems.

EQUIPMENT REQUIRED

Resistors

3 — 250-Ω (200-W or more), wire-wound

3 — 100-Ω, 33-kΩ

Instruments

1 — DMM

2 — 1000-W-maximum wattmeters (HPF)

Power Supply

Three-phase, 120/208-V, 60-Hz

EQUIPMENT ISSUED

TABLE 15.1

Item	Manufacturer and Model No.	Laboratory Serial No.
DMM		
Wattmeter		
Wattmeter		

Since more than one resistor of a particular value will appear in some configurations of this experiment, the measured value of each should be determined just prior to inserting them in the network.

RÉSUMÉ OF THEORY

In a balanced three-phase system, the three-phase voltages are equal in magnitude and 120° out of phase. The same holds true for the line voltages, phase currents, and line currents. The system is balanced if the impedance of each phase is the same. For the balanced Y load, the line current is equal to the phase current, and the line voltage is the square root of 3 (or 1.73) times the phase voltage. For the balanced Δ load, the line and phase voltages are equal, but the line current is the square root of 3 (or 1.73) times the phase current.

The relationship used to convert a balanced Δ load to a balanced Y load is

$$\boxed{\mathbf{Z}_Y = \frac{\mathbf{Z}_\Delta}{3}} \tag{15.1}$$

Since the sum (phasor) of three equal quantities 120° out of phase is zero, if a neutral wire is connected to a balanced three-phase, Y-connected load, the current in this neutral wire will be zero. If the loads are unbalanced, the current will not be zero.

The phase sequence of voltage does not affect the magnitude of currents in a balanced system. However, if the load is unbalanced, changing the phase sequence will change the magnitude of the currents. The sequence of voltages determines the direction of rotation of a three-phase motor.

The total average power to a three-phase load circuit is given by Eqs. (15.2) and (15.3):

$$\boxed{P_T = 3E_p I_p \cos \theta} \tag{15.2}$$

where E_p is the phase voltage, I_p the phase current, $\cos \theta$ the power factor, and θ the phase angle between E_p and I_p.

$$\boxed{P_T = \sqrt{3} E_L I_L \cos \theta} \tag{15.3}$$

where E_L is the line voltage, I_L the line current, $\cos \theta$ the power factor, and θ the phase angle between E_p and I_p.

The power to a three-phase load circuit can be measured in two ways:

1. Three-wattmeter method
2. Two-wattmeter method

THREE-WATTMETER METHOD

The connections for the wattmeters for a Δ and a Y circuit are shown in Fig. 15.1. The total power P_T is equal to the sum of the wattmeter readings.

Note that in each case, connections must be made to the individual load in each phase, which is sometimes awkward, if not impossible. A more feasible and economical method is the two-wattmeter method.

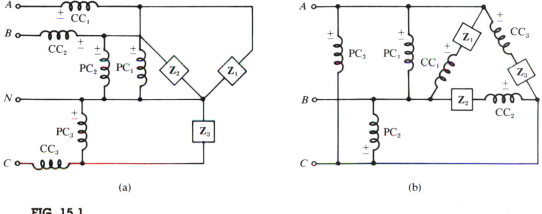

(a)   (b)

FIG. 15.1

TWO-WATTMETER METHOD

With the two-wattmeter method, it is not necessary to be concerned with the accessibility of the load. The two wattmeters are connected as shown in Fig. 15.2.

It can be shown that the two-wattmeter method of measuring total average power yields the correct value under any conditions of load.

If we plot a curve of power factor versus the ratio of the wattmeter readings P_1/P_2, where P_1 is always the smaller reading, we obtain the curve appearing in Fig. 15.3.

The graph shows that if the power factor is equal to 1 (resistive circuit), then $P_1 = P_2$ and the readings are additive. If the power factor is zero (reactive), then $P_1 = P_2$; but P_1 is negative, so $P_T = 0$. At a 0.5 power factor, one wattmeter reads zero, and the other reads the total power. It is indicative from the graph that if $F_p > 0.5$, the readings are to be added; and if $F_p < 0.5$, the readings are to be subtracted.

The two-wattmeter method will read the total power in a three-wire system even if the system is unbalanced. For three-phase, four-wire systems (balanced or unbalanced), three wattmeters are needed.

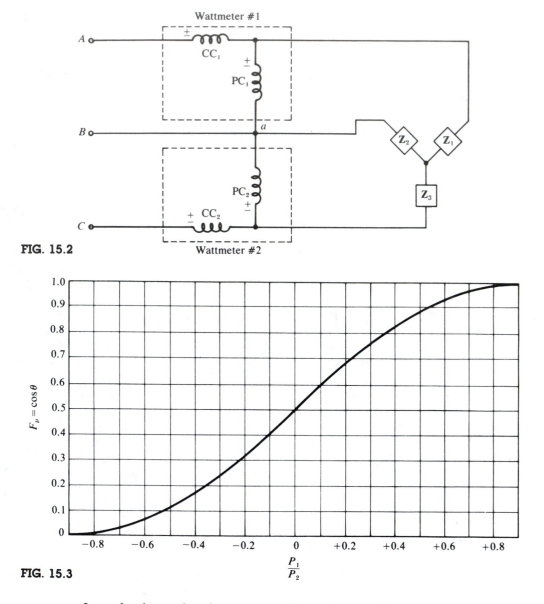

FIG. 15.2

FIG. 15.3

It can be shown that the power read by each wattmeter is given by the following equations:

$$P_2 = E_L I_L \cos(\theta - 30°)$$

(15.4)

$$P_1 = E_L I_L \cos(\theta + 30°)$$

(15.5)

$$P_T = P_2 \pm P_1$$ (the algebraic sum of P_1 and P_2, where P_1 is the smaller reading)

(15.6)

Another advantage of the two-wattmeter method is that it allows us to calculate the phase angle (θ) and therefore the power factor (F_p) using the expressions

$$\tan \theta = \sqrt{3}\,\frac{P_2 - P_1}{P_2 + P_1} \qquad\qquad\text{(15.7)}$$

$$F_p = \cos \theta \qquad\qquad\text{(15.8)}$$

PROCEDURE FOR CONNECTING THE WATTMETER USING THE TWO-WATTMETER METHOD

Connect the potential coils (PC) in the circuit last, and use the insulated alligator clip leads for their connections to line *B*. (See Fig. 15.2.) Close the breaker. If either wattmeter reads down-scale, shut down and reverse the current coil (CC) of the wattmeter that reads down-scale. Apply power again and read both wattmeters.

To find the sign of the lower-reading wattmeter, proceed as follows. If wattmeter #1 is the lower reading, unclip lead *a* of the PC coil and touch to any point on line *A*. If the wattmeter deflects up-scale, the reading is to be added (+). If the deflection is down-scale, the reading is negative and is to be subtracted (−). If wattmeter #2 is the lower-reading wattmeter, then the PC lead *a* is touched to line *C* for the test.

PROCEDURE

Note: Be absolutely sure that the power is turned off when making changes in the network configuration or inserting and removing instruments.

Part 1   Resistors in a Balanced Δ

Construct the network of Fig. 15.4

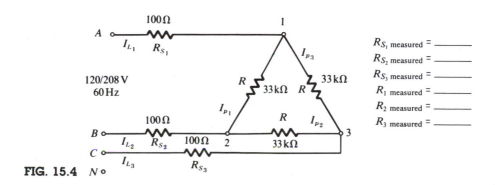

FIG. 15.4

(a) Measure the three-phase voltages and line voltages.

$V_{AB} =$ _____ , $V_{BC} =$ _____ , $V_{CA} =$ _____

$V_{12} =$ _____ , $V_{23} =$ _____ , $V_{31} =$ _____

What is the relationship between the line voltage and the phase voltage in a balanced Δ-connected load?

(b) Measure the voltage across the sensing resistors with the DMM.

$V_{R_{S_1}} =$ _____ , $V_{R_{S_2}} =$ _____ , $V_{R_{S_3}} =$ _____

Calculate the line currents using the sensing voltages and Ohm's Law.

$I_{R_{S_1}} = I_{L_1} =$ _____ , $I_{R_{S_2}} = I_{L_2} =$ _____ ,

$I_{R_{S_3}} = I_{L_3} =$ _____

(c) Using Ohm's Law, and the measurements of part 1(a), calculate the phase currents.

$I_{p_1} =$ _____ , $I_{p_2} =$ _____ , $I_{p_3} =$ _____

(d) For a balanced Δ-connected load, how are the line currents related (in magnitude)?

(e) For a balanced Δ-connected load, how are the phase currents related (in magnitude)?

(f) For a balanced Δ-connected load, how are the line and phase currents related (in magnitude)?

Part 2   Balanced Four-Wire, Y-Connected Resistor Load

(a) Construct the circuit of Fig. 15.5. Refer to the procedure for connecting the wattmeters. *Exercise caution at all times!* If in doubt, ask your instructor. Insert the measured value of each resistor.

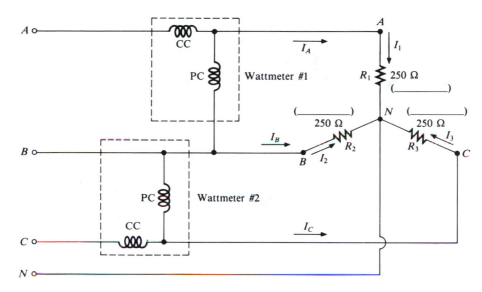

FIG. 15.5

Measure the voltages V_{AN}, V_{BN}, and V_{CN}.

$V_{AN} =$ _____ , $V_{BN} =$ _____ , $V_{CN} =$ _____

Are these the line or phase voltages? Explain.

Measure the voltages E_{AB}, E_{BC}, and E_{AC}.

$E_{AB} =$ _____ , $E_{BC} =$ _____ , $E_{AC} =$ _____

What is the relationship between (V_{AN}, V_{BN}, and V_{CN}) and (E_{AB}, E_{BC}, and E_{AC})?

Calculate E_{AB}/V_{AN}, E_{BC}/V_{BN}, and E_{AC}/V_{CN}.

$E_{AB}/V_{AN} =$ _____ , $E_{BC}/V_{BN} =$ _____ ,

$E_{AC}/V_{CN} =$ _____

Do these ratios confirm the theory? Explain.

(b) Calculate the magnitudes of the currents I_1, I_2, and I_3.

$I_1 =$ _____ , $I_2 =$ _____ , $I_3 =$ _____

Are these the line or phase currents? Explain.

How are these currents related to the magnitudes of I_A, I_B, and I_C?

Calculate the magnitudes of I_A, I_B, and I_C.

$I_A =$ _____ , $I_B =$ _____ , $I_C =$ _____

(c) Calculate the power delivered to the load.

$P_T =$ _____

Read the wattmeters.

$P_1 =$ _____ , $P_2 =$ _____

Calculate P_T from P_1 and P_2.

$P_T =$ _____

(d) Calculate the power factor F_p and the phase angle θ.

$F_p =$ _____ , $\theta =$ _____